D1045194

— THE —
ANGLER'S FLY
IDENTIFIER

THE ANGLER'S FLY IDENTIFIER

The Complete Guide to Insects and Artificials

DR. STEPHEN J. SIMPSON &
DR. GEORGE C. MCGAVIN

RUNNING PRESS
PHILADELPHIA · LONDON

A QUINTET BOOK
Copyright © 1996 Quintet Publishing Limited.

9 8 7 6 5 4 3 2

Digit on the right indicates the number
of this printing

ISBN 1–56138–610–3

Library of Congress
Cataloging-in-Publication Number
94–74907

This book was designed and produced by
Quintet Publishing Limited
6 Blundell Street
London N7 9BH

Creative Director: Richard Dewing
Designer: Ian Hunt
Project Editor: Diana Steedman
Editor: Tony Hall
Photographers: Jeremy Thomas, Peter Gathercole

Typeset in Great Britain by
Central Southern Typesetters, Eastbourne
Manufactured in Hong Kong by Regent Publishing Services Ltd
Printed in China by Leefung-Asco Printers Ltd

Running Press Book Publishers
125 South Twenty-second Street
Philadelphia, Pennsylvania 19103-4399

CONTENTS

INTRODUCTION

Fly fishing for trout is a remarkable sport, combining grace in technique with powers of observation and interpretation possessed by field naturalists. Or at least that is what fly fishing can be when practiced by an expert. The problem is that the mystique and jargon, which obscures the subject, often discourage anglers from seeking these ideals.

The fundamentals of fly fishing are easily and relatively inexpensively acquired. The realization that this is the case has led to an explosion in interest in the sport. Many of us nowadays learn to fish by casting and retrieving lures which, as far as we can tell, bear no obvious resemblance to any particular natural food item. Perhaps to a trout they do, but that is not the point: the angler has not selected the fly to match a natural organism. Lure fishing is certainly fun, and often productive, but it omits a vital piece of the intellectual puzzle: the identity and behavior of the trout's prey.

Herein lies one of the great impediments to anglers moving beyond lure fishing. Trout feed on a huge range of animals, most of them insects. There are literally hundreds of thousands of types (species) of insect that may end up in the stomach of a trout, let alone all the other things like snails, shrimps, tadpoles, and fish. Most of these insects have life cycles during which they completely change not only in size but also in form and behavior. Worse still, each creature is blessed with an arcane Latin or Greek name as well as any number of locally relevant common names. Finally, to top it all off, the artificial flies that have been tied to imitate some of these animals have their own, often uninformative titles.

In short, working out that the trout are feeding on *Boojumas snarkus* (variously known, according to locale, as the Gray Ghost Thumper, the Waddlepooper and the Mudfangler), and relating this to appropriate fly patterns such as Carroll's Revenge, or The Hoax, is not made easy. But it can be. That is the point of this book.

Trout fishing combines dexterity in technique with powers of observation possessed by the best field naturalists.

PART I
THE TROUT
AND ITS PREY

IDENTIFYING FOOD ANIMALS THE EASY WAY: AND WHY BOTHER?

The key to success is to ignore all but essential jargon. While it is true that trout eat hundreds of thousands of different animal species, these are members of a much smaller number of basic groups. In fact, there is no need to learn to identify more than a couple of dozen types of animal. Identifying them down to species provides little useful extra information, beyond that which is obvious by looking at the creature. Once the group to which an animal belongs has been established, then the selection of an appropriate pattern of artificial fly can be made by reference to features such as the animal's size, form, and color.

Phryganea grandis *or, more simply, a large brown caddisfly.*

Perhaps trout are feeding on adults of *Phryganea grandis.* Having identified the insect as a member of the order Trichoptera (commonly known as caddisflies or sedges), then the choice of fly would be a sedge pattern, colored reddish brown and tied on a number 8 hook. There are literally thousands of types of caddisflies worldwide: all you need to know is that the insect is a large, brown species.

Caddisfly artificial: Large Brown Sedge.

A selection of typical river dry flies.

By this reasoning you might argue, why bother even identifying an insect to a major group? Why not just choose from your fly box the nearest fly in appearance to the real thing? In certain instances this approach can be effective, but it relies on your being able to see the food organism. In many instances, however, you can only infer what the trout are feeding on by other, indirect clues. It might be, for example, that trout are feeding on something near the surface of the water. There is no sign of fish taking winged insects, although there are adult caddisflies darting around the margins. It would, therefore, be worth considering fishing an emerging caddis pupa. Knowing that the winged insects are in fact caddisflies makes the choice of a pattern representing their pupa a simple and informed operation. Not only that, but knowing the identity of the insect allows you to present the artificial fly realistically. This is because names provide pass-words to sources of information about natural history and behavior. Once acquired, this information becomes an integral part of the identification.

There are obviously direct, practical advantages to being able to identify food organisms. There are also indirect benefits. Simply knowing the names of the animals (and plants) around you increases the pleasures of being by the waterside. It heightens your powers of observation and adds confidence. As a result, fly fishing loses a great deal of its mystique, yet none of its pleasure.

11

THE BASIC PRINCIPLES OF BIOLOGICAL CLASSIFICATION

Organisms are classified by scientists in a hierarchical manner, being grouped into larger and larger categories. The most fundamental unit of classification is the species. Although the species concept is not quite as clear-cut as we biologists would like, it is broadly true to define a species as a unit made up of individuals which are genetically compatible,

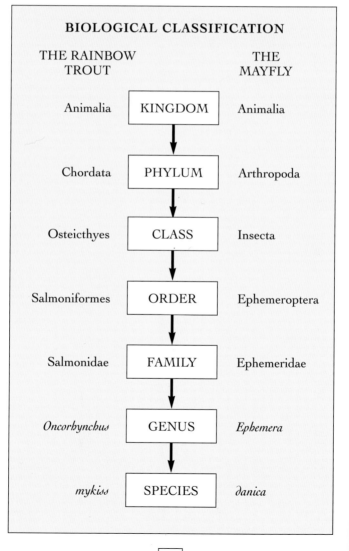

BIOLOGICAL CLASSIFICATION

THE RAINBOW TROUT		THE MAYFLY
Animalia	**KINGDOM**	Animalia
Chordata	**PHYLUM**	Arthropoda
Osteicthyes	**CLASS**	Insecta
Salmoniformes	**ORDER**	Ephemeroptera
Salmonidae	**FAMILY**	Ephemeridae
Oncorhynchus	**GENUS**	*Ephemera*
mykiss	**SPECIES**	*danica*

A brace of Brown Trout.

13

i.e. able to interbreed and produce reproductively viable off-spring. Species are combined into the next most fundamental grouping: genera. In fact, the Latin or Greek name that graces each described organism is made up of two parts, the genus and species. Hence, the Rainbow Trout is *Oncorhynchus* (the genus) *mykiss* (the species within the genus). Perhaps surprisingly, not all the animal species that are alive today have been described and named. In fact, most have not even been discovered, and most of these are insects.

Genera are next grouped into families, families into orders, orders into classes, classes into phyla, and phyla into kingdoms. Within each of these there are also various subdivisions (e.g. superfamilies are groups of families within an order, while subfamilies are groups of genera within a family). So, the Rainbow Trout is Kingdom Animalia (including all multicelled organisms that ingest food), Phylum Chordata, Subphylum Vertebrata (all animals with backbones), Class Osteichthyes (all bony fish), Order Salmoniformes (all bony fish with a single dorsal fin placed, like the pelvic fins, at least halfway along the body), Family Salmonidae (all trout, salmon, char, grayling, and others that possess an adipose fin between the dorsal and tail fins), Genus *Oncorhynchus* (all Pacific trout and salmon), species *mykiss*.

As a result of advances in the science of classification (taxonomy or, more fashionably, systematics), biologists quite regularly change their mind about the relationships between species. Most agree that the aim in classifying organisms is not just to catalog them, but to reflect the order in which they evolved. Ideally, the pattern of classification should represent the branching tree of evolution.

When new evidence, or a reassessment of that which already exists, leads to revision of a classifactory group, then this will often result in the renaming of species. The Rainbow Trout, for instance, was called *Salmo gairdneri* (i.e. was considered to be a member of the same genus as the Brown Trout and the Atlantic Salmon) until 1989, when it was accepted that it better warranted inclusion in the genus *Oncorhynchus*, which includes the various species of Pacific salmon and trout. Actually, the taxonomy of trout and salmon is among the most contentious of any animal group.

HOW TO USE THIS BOOK

This book is a practical guide to insect identification. Brief biographical sketches serve to introduce the reader to the fascinating lives of the various prey insects. Such background information provides a biological context to enable the angler to interpret what the insects and other prey animals are actually doing.

Another crucial aspect of the story is the trout itself. A short section is provided on relevant aspects of trout behavior. The theme here, as elsewhere in the book, is that the angler is, in effect, an experimental scientist, whose aim is to second-guess the behavior of the fish and to discover the

Descriptions of prey insects precede the artificial fly patterns.

key stimuli that will release a feeding response toward a piece of sharp, barbed wire wrapped in fur, feathers, and tinsel. We introduce some of the concepts used by professional biologists who study animal behavior. Such ideas provide a framework within which the angler can interpret his or her own experiences of trout behavior.

Familiarity with the trout's natural habitat is vital to understanding its behavior.

The forward cast.

We have tried to write the book in such a way as to transcend the geographic region and type of water being fished, (whether that be a small clear-water lake, a large reservoir, a lowland chalk stream, or a mountain stream), and the species of trout being sought. As a result, the book may err at times on the side of over-generalization, but the basic principles ought to be translatable into more specific situations.

The text leads in a logical progression from observation to deduction, then from identification to biological background and fly patterns.

STEP ONE: observing and deducing. Begin at the "decision key" on page 32. This is a simple series of questions that good trout anglers ask themselves as they sit observing the water before starting to fish. Careful observation provides the evidence needed to make successful decisions about how and where to fish.

River fishing with a dry fly in early summer.

Wading the stream.

STEP TWO: **identifying the food animals.** Turn to the identification key on pages 40–43. During observation and deduction it will often be necessary to identify an animal that evidence or logic suggests is trout fodder. Hence, the decision key leads at appropriate points to the identification key. Having established to which group the animal belongs, turn to the appropriate section in the Identifier to learn more about your specimen.

STEP THREE: choosing and presenting an artificial fly.
Once the identity of a food animal has been established,
choose an artificial from the selection of fly patterns that
represent the natural flies or organisms. The fly patterns
follow the natural insects within their group in the sequence
of their life cycle: e.g. for mayflies, nymph, sub-imago (dun),
adult (spinner), dead (spent).

We have omitted from our selection of fly patterns a vast
number of valid imitations, but that does not matter. Of far
more importance than listing all the myriad variations on a
theme is to provide some information about how best to
present the fly to the trout. Understanding how to do this involves knowing something about the behavior and lifestyle of
the food animals and of the trout.

Selecting an appropriate artificial fly.

The Principles of Trout
Feeding Behavior

Trying to fish for trout without understanding their behavior is about as futile an occupation as brain surgery with elementary woodworking skills.

Trout are complex animals and their behavior is the result of a vast array of stimuli coming both from the external environment and from within the animal itself. These stimuli are collected and transferred to the brain by sense organs tuned to receive stimuli which are most relevant to the trout's lifestyle and ecology. The sense organs, and the physiological control systems which receive their information, have been fine-tuned by millions of years of evolution. As a result, trout are equipped not only to cope with the basic features of a wide range of aquatic environments but also to respond successfully in the face of the regular changes imposed by their world.

STIMULI, ETHOLOGY, AND DUTCH MAIL VANS

The three main types of stimuli coming from the external environment are visual (color, form, and movement), mechanical (touch, sound, and vibration) and chemical (taste and smell). Such stimuli may produce, among other responses, one of four major classes of behavior: feeding, escape, aggression, and sexual activity. These in turn are made up of behavioral subcomponents, each of which has a stimulus threshold which must be exceeded before the behavior will occur. The stimuli which are particularly effective in crossing these thresholds are known as *sign stimuli*.

Trout anglers are, or should be, students of animal behavior. The academic study of animal behavior, with particular reference to the relationship between the animal and its natural environment, is called ethology. One of the prime aims of the ethologist is to discover the nature of the sign stimuli that elicit a particular behavior. That is exactly what the trout angler aims to do when attempting to entice a fish to take an artificial fly.

A male stickleback in mating colors courts an egg-laden female.

One of the fathers of ethology was the Dutch scientist Niko Tinbergen, who, in 1973, received the Nobel Prize for his contributions to biology. He used to great effect the technique of making artificial models to help define the sign stimuli eliciting behavior.

Tinbergen discovered that the red belly of sexually mature sticklebacks is the cue that elicits aggression in rival males. He deduced this when the male fish in his aquarium in Leiden shot across to the side of the tank nearest the window and rammed the glass every time the red mail van drove by outside. Subsequent experiments using wooden model fish indicated that the red belly was of far more importance than any other visual feature. A model could be a perfect mimic of a male stickleback in all respects bar the belly and receive no attacks, yet if a totally unfishlike object was painted red on its lower half the sticklebacks responded vigorously.

There is an important message here for us trout anglers: we do not perceive the world as a trout does, because both our sensory organs and the way our brains interpret their information differ. What to us may appear to be a perfect replica of a natural insect may totally lack the critical ingredient for a trout. Similarly, something may look nothing like a real prey animal to us, yet bear all the key sign stimuli for a trout. Indeed, Tinbergen and many others since, have shown that you can unnaturally exaggerate certain sign stimuli and produce an even greater response than the real thing. Such models provide what are termed *supernormal stimuli*. The success of many trout flies and lures is undoubtedly due to the fact that they represent "caricatures," which elicit super-normal reactions.

THRESHOLDS OF BEHAVIOR

Behavioral thresholds are embodied in patterns of neural activity in the brain. The nature and intensity of external stimulation required to exceed a behavioral threshold is not fixed. It varies with factors such as time of day or stage in the season, when the fish last fed and its stage of sexual development, its previous experience, the temperature and oxygenation of the water. As a result, a trout will not always respond in the same way to a particular external stimulus. Hence a fly pattern which proved deadly one day need not evoke any interest the next.

There is a famous adage in the academic study of animal behavior which is particularly apt for trout. It is called the Harvard Law of Animal Behavior, and goes as follows: "Under sufficiently well controlled conditions an animal chooses to do as it damn well pleases." Use this as solace when all else fails.

THE FEEDING SEQUENCE

Feeding is the trout behavior that is of most direct concern to an angler. While it is widely believed that aggressive responses can lead to trout taking artificial flies, particularly garishly colored lures, there is no firm evidence that this is true, nor that trout take such offerings due to inquisitiveness or playfulness. It is clear, however, that artificial flies and lures provide sign stimuli that cause a trout both to orient toward them and to take them into its mouth.

Blue-winged Olive Thorax.

Feeding consists of a sequence of behavioral elements. First the fish orient toward a potential food item. Next it moves so that the prey is directly in front of its face. Then it opens its jaws and sucks the object into its mouth. Finally, it swallows the prey. Each stage is triggered by external stimuli, and the sequence can be aborted at any point if the stimuli provided by the potential meal are inappropriate.

Orientation toward the prey. The two main cues that elicit orientation are visual and vibrational, the balance of the two varying according to light intensity and water clarity. Additionally, trout have a keen sense of smell and can detect minute quantities of food particles dissolved in the water.

A Brown Trout uses its excellent vision to locate prey.

Visual stimuli provided by prey fall into three main categories: shape, color, and movement. The combined power of these three to stimulate the trout is greater than the sum of their individual contributions. When all three tally, they provide strong inducement.

Trout have excellent eyesight and see color, although this does not mean that they see colors in the way that we do. They possess visual receptor cells (called cones) in their retina which are specially tuned to red, blue, and green light. Additionally, young trout of up to a year or two old have another receptor type which is sensitive to ultraviolet light. As the fish ages, it loses the ability to sense ultraviolet light. One other interesting feature is that the peak sensitivity of the red cones varies throughout the season, moving toward shorter wavelength during summer. Perhaps this explains the common observation by anglers that some species of trout (rainbows especially) respond best to orange flies early and late in the season.

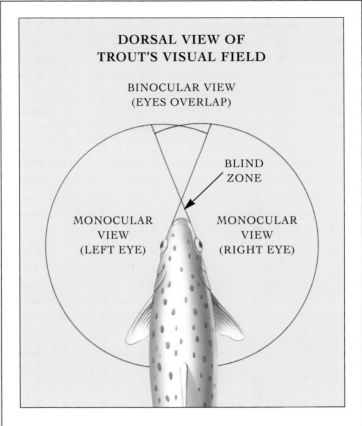

DORSAL VIEW OF TROUT'S VISUAL FIELD

BINOCULAR VIEW
(EYES OVERLAP)

BLIND ZONE

MONOCULAR VIEW
(LEFT EYE)

MONOCULAR VIEW
(RIGHT EYE)

Trout eyes each possess a spherical lens with a short focal length. Scattering and absorption of light by suspended particles and dissolved material severely degrades the quality of an image over distances of more than a few yards under water, so trout are in effect rather nearsighted.

Except for blind spots to the rear and in front of the face, trout have an almost 360° view of the world. Most parts of the visual field are seen with only one of the eyes, but there is a conical region of about 45° in front of the trout's head where the fields of view of the eyes overlap. When a hungry trout sees a prey in the periphery of its vision it turns toward the object, thus bringing it into the binocular field. The trout then moves toward the quarry and tracks its movements. If the prey changes direction or speeds up, so too does the trout. As the fish homes in, the prey expands in the visual field and come into closer focus, and thus scrutiny.

If, during the chase, features of the prey's appearance do not continue to convince the trout's brain that the object really is edible, then the fish either veers away or decelerates. This is why trout are hard to convince when the water is still and clear, the light good and the current slow, and also why fish are often less choosy when standing station in a fast current, where food rushes by providing little time for discrimination.

Vibrational stimuli are also used by trout to orient toward prey, and are particularly relied upon when visibility is poor. Fish have two organs for detecting water-borne vibrations: the ears (which respond to high-frequency vibrations, i.e. sound, and are located on either side of the head) and the lateral line sensors (which detect low-frequency vibrations and are visible as a line running along each flank of the fish). Using both types of highly sensitive organ a trout is able to detect and orient toward small prey over greater distances than it can see, even when visibility is good. Similarly, the sound of a poorly presented fly and line splashing on the surface, or of a size ten rubber boot crunching on boulders will cause the fish to dart for cover.

The attack. When the quarry reaches a critical distance from the face of the fish, the trout opens its jaws and engulfs it. Rapid acceleration of a fleeing prey induces a burst of speed by the trout and a savage attack. This is the basis for the success of the "induced take" technique, and why a trout that has followed an artificial fly throughout the retrieve may then

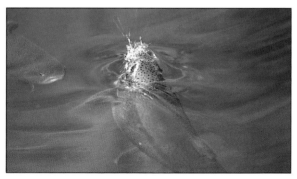

A Rainbow Trout taking a dry fly.

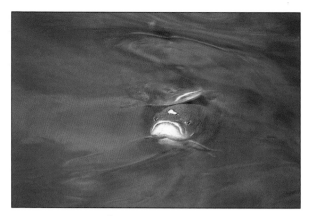

Two Rainbow Trout vying for a cast fly.

slash at the imitation just as the angler lifts the rod in readiness for the next cast.

If a trout repeatedly follows an artificial fly without taking it, the reason is that the fly has provided sufficient stimulation to trigger movement toward it, but fails to exceed the trout's behavioral threshold for the final attack. Perhaps this is because of the color, shape, movement, or smell of the artificial, or because the trout has a high threshold for feeding due to other internal or environmental factors.

Swallowing. Once the object has entered the trout's mouth, other sensory qualities come into play, namely texture and taste. The mouth is richly endowed with sensory cells which are linked to the swallowing and ejection reflexes. This explains why artificial flies are often taken into the mouth but then rapidly expelled.

THE TROUT'S SELECTIVE "ATTENTION"

At any given time a trout will be faced with a veritable cafeteria of available prey, with the items on the menu changing from hour to hour during the day and from day to day throughout the season.

Many studies have shown that predators respond like economists in the marketplace, choosing their food items to maximize their net rate of energy gain. Without knowing it, they trade-off the costs and benefits associated with various feeding options and choose the most profitable prey types.

There are four basic variables in the cost-benefit analysis: (1) the animal's nutritional requirements at that time, (2) the abundance of different prey types available in the environment, (3) the nutritional value of these prey, and (4) the costs (including energy, time, and increased risk of predation) associated with their capture and processing.

The first and last of these variables alter with factors such as the trout's age, size, and reproductive state, the time of day, season, and, particularly important for cold-blooded animals such as trout, temperature. The bigger you are, the less the energetic costs of chasing small prey are justified by the nutritional returns. Basically, it is only worth eating small things rather than bigger prey when there are lots more of them and they are cheap to acquire: otherwise it is more profitable to eat fewer large prey. This explains why very large trout are more likely to feed on bigger prey such as fish, frogs and crayfish than on small insects.

When the water is coldest (e.g. during winter, early spring, and late autumn) the energetic rates and requirements of trout are low and it pays them either not to feed at all, or only to accept especially large prey that happen past. As a result, it is easier to tempt trout at these times with large lures than with small insect imitations.

When there is a large hatch of winged insects (or any other superabundance of prey) it benefits the trout to concentrate its feeding on those, for the feast is both easily gained and short-lived. Under these circumstances, trout become responsive to stimuli provided by the commonest prey species and seemingly ignore all others. Such a response probably involves the trout's learning, through recent feeding experience, a set of stimuli possessed by the target prey and ignoring all others. The result is that, for a period, the trout will only accept artificial flies that represent the hatching insect. The memory is short-lived in the brain, lasting for only as long as that particular prey type is available. Unfortunately for the angler, it may mean that, with such an abundance of the real thing available, the fish are particularly discriminating and hard to convince to take an artificial fly.

A large Brown Trout which has taken a dry fly.

LEARNING BY EXPERIENCE: THE CLEVER TROUT

Although they are not as clever as many anglers would like to think, trout do exhibit a variety of forms of learning, as well as short- and long-term memory. They demonstrate, and use in their feeding behavior, both *nonassociative* and *associative learning*. The former involves two common responses: *habituation* and *sensitization*.

Habituation is where an animal starts to ignore a stimulus with repeated presentation. It is common for a trout to follow an artificial fly for the first few times it is presented, but when the angler, encouraged by the trout's response, continues to cast the same pattern and retrieves it in a similar manner, the fish soon ignores it. The trout's brain has devalued the stimuli provided by the fly, such that the orientation threshold is no longer crossed by the stimuli provided. *Dishabituation* occurs when the devalued stimulus is changed, and the subsequent response may actually be more vigorous than it would have been without prior habituation. This explains why when after a period of fruitlessly casting a given pattern of fly, an angler often meets success on the first cast after the pattern, color, or retrieve rate is changed.

Sensitization is where the response, rather than waning, becomes more pronounced with repeated stimulation. Unfortunately for anglers, this is a process often seen in the context of the trout escape response. Once spooked, a trout becomes "edgy" and highly responsive, interpreting even the slightest splash or movement as an approaching enemy. Such an agitated state can last for hours afterward.

Associative learning is where the animal learns to associate a stimulus, or combination of stimuli, with an effect. Trout, like all animals, enter the world with a set of preexisting responses to stimuli. On top of this, however, they are equipped with the ability to learn to respond to the "predictably unpredictable."

Given the range of aquatic environments in which they live and the breadth of food organisms they eat, there is no way that a trout could be preconditioned at birth to respond to all the various specific combinations of prey and other stimuli that it might encounter during its life: even though it is predictable that such variation will occur and will need to be coped with. Hence, trout, like all animals, can learn a variety of ecologically relevant stimulus associations.

Trout can learn to associate visual, olfactory (smell), and mechanical stimuli with food or enemies. This is known as *classical* or *Pavlovian conditioning*, and is almost certainly a central mechanism guiding trout foraging. It also explains why trout that have been hooked before are much harder to convince to take an artificial offering than are naive fish: they have learned an association between features of the artificial fly (or perhaps the leader and line) and the unpleasant sensation of being hooked and pulled toward the bank.

Trout can also learn from the consequences of their actions, a process termed *operant* or *trial and error learning*. This capability helps the animal adjust to its surroundings and learn how best to locate food. It is probably vital to stocked fish when first settling in to their new home, and also helps established trout cope with the inevitable changes in their world which occur as the seasons and years progress.

Other facets of "intelligence" possessed by trout are the ability to learn to navigate within their environment, and to associate food and feeding sites with the time of day (as measured by their biological clock). Navigation involves, among other things, another form of learning called *latent learning*, whereby the trout recognizes key visual and olfactory features of its environment.

EQUIPMENT FOR DISCOVERING THE TROUT'S PREY

There are several accessories that would prove far more useful additions to the kit of a trout angler than many of the other accoutrements that provide powerful stimuli for the wallet-emptying reflex of trout anglers, but do little for the trout.

Sweep ("butterfly") net. This is a vital piece of equipment needed for catching winged insects. A sweep net is simple and inexpensive to make. (1) Take a wire coat hanger, straighten the hook, and bend the triangular section into a loop. (2) Bind or tape what was the hook onto the end of 3–6 ft bamboo garden cane. (3) Take a piece of mesh (mosquito netting or muslin) and sew a cone with a mouth the diameter of the circular wire rim and a depth approximately 1½–2 times the diameter of the opening. Make it deeper rather than shallower, so that insects caught in the bottom of the net can be trapped by rotating the handle through 90°, causing the mouth to be closed as the body of the net flops across it. (4) To attach the net to the rim, simply fold and sew a hem over the wire.

Pond net. Just as a sweep net is needed to catch winged insects, a pond net is essential for sampling life beneath the water surface. Again, these are cheap to buy or make. Visit your regular supplier of beach balls, plastic pails and shovels, and purchase one of those fetching nets that children use for dredging rock pools. Make sure it has a suitably fine mesh. Alternatively, but no better for riverside credibility, attach a wire cooking sieve (about 8 in diameter) to a 6 ft length of garden cane.

Jam jar, forceps, and a white plastic tray. More high-tech gear. Jam jars filled with pond water are an excellent means of identifying and watching the fruits of pond netting. Take home whatever you cannot identify at the time, for later study. Try not to put too many species in one jar, otherwise you will be left with a small number of predatory organisms, albeit contented ones.

The plastic tray comes in handy for spreading out the contents of a pond sweep, prior to transferral to the jam jar with the forceps.

A useful tray can be had from an empty 2 pint, rectangular plastic ice- cream container. Keep your lunch in it up to the point when the tray is needed for science.

Hand lens. This is a particularly useful item for closely observing small insects and crustaceans.

Binoculars. These are extremely helpful aids when trout are feeding on hatching insects which are out on the water and not available to your sweep net. If they won't come to you, you can get closer to them with binoculars, and perhaps be able to see enough to make an accurate identification.

Polaroid glasses. Most anglers have these as standard accessories. So they should, both for eye protection against flying hooks and for making observing trout under the water easier by reducing surface glare and reflection.

FISHING DECISIONS

I n this chapter we provide a logical series of questions that the angler should ask before making the first cast of the day. Don't stop there, however: once fishing has commenced it is vital to continue to ask the same questions throughout the day. The angler must remain alert to changes in the behavior of the fish and of their prey.

The assumption is made that you are alone at the waterside. If not, then make use of another useful source of information: fellow sufferers. Ask other anglers what has, and has not, worked for them. Similarly, if there is one, talk to the fishery manager and read the recent entries in the logbook detailing catch returns and successful fly patterns.

The sequence of questions is summarized on pages 32–33, and elaborated upon as follows.

Reading the water is an important skill.

THE DECISION SPREAD

Start here → **1** Is there evidence of trout feeding at or near the surface?
YES   NO

2 Are trout taking winged insects from the water surface?
YES   NO

Catch, identify and fish appropriate fly pattern.
Go to pages 40-43.

3 Are trout slashing at the surface?
YES   NO

Probably feeding on small fish. Try imitating escaping or dead, floating fish.
Go to pages 184-185.

4 Are trout browsing on the surface, showing their dorsal and tail fins, and sipping the surface film (i.e. head and tail rises)?
YES   NO

Probably feeding on midge pupae (buzzers). Sample with pond net, look for empty "shucks", and catch any adult midges. Choose and fish appropriate pattern.
Go to pages 124-128.

You have missed something—start again!

5 Can you see the trout, even though they are not feeding on the surface?
YES   NO

Observe carefully and deduce where and on what they are feeding. Identify pond and sweep net samples to guide your choice of fly.
See pages 40-43.

6 Are there winged adult insects apparent, even though trout are not taking them from the surface?
YES   NO

Catch and identify them and, if aquatic, fish their immature forms in likely lies. Read the water to locate these.
See pages 40-43.

7 Take samples with the pond net, choose likely lies and systematically fish imitations of species collected, starting with the most abundant species.

Still no luck? Try systematic lure fishing.
Go to pages 38–39.

Still no trout?
Seek consolation.

QUESTION 1: **Is there evidence of trout feeding at or near the surface? If yes, continue to question 2. If no, go to question 5.**

Surface feeding activity is characteristic of trout taking particular types of prey. You may need to learn to see surface feeding fish, since clues can be as subtle as the flattening of a ripple or a barely perceptible bow wave.

QUESTION 2: **Are trout taking winged insects from the water surface? If yes, continue. If no, go to question 3.**

If winged insects can be seen being taken from the surface then life is made relatively easy. What you need to do is capture one of the insects, or if that is not possible, try to identify them from a distance. It is also worth inspecting the vegetation and rocks around the margins of the water. Look for dead adults washed into the margins. Once you have captured what the trout are eating, then turn to the identification key on pages 40–43.

The behavior of the food insects and of the feeding trout gives crucial clues as to how to present your artificial fly. There are several ways in which trout feed on winged insects at the surface. The commonest is the rise in which a resting insect is sucked down as it floats along on the current. The result is a series of concentric ripples spreading outward from where the insect was sitting before embarking on a tour of the trout's alimentary canal.

This form of rise is often seen for newly emerged mayfly duns, as they sail sedately along prior to taking to the air for the first time. Under such circumstances the trout are in no hurry. They have plenty of time to observe the insect. It is crucial that the artificial fly sits realistically on the water and floats naturally with the current.

Similar forms of rise occur when trout scavenge on dead insects floating, waterlogged in or just below the surface film. Here the trout has an even better view of the prey. Obviously, no movement other than that from the current should be imparted to the artificial fly.

On other occasions the fish are less leisurely in their capture of winged insects. Slashing rises at the surface, leaps to take airborne insects, and underwater swirls and bow-waves are all indicators that there is furious activity going on beneath the surface. Usually such behavior is associated with prey that are moving quickly and likely to escape.

A typical case is when caddisflies are emerging. When fully developed, the pupae rise rapidly from the safety of the river or lake bed and emerge from their pupal cuticle as winged adults, in some species with all the speed and alacrity of a quick-change artist in a stage farce. They then scoot across the water and launch themselves into the air. All of this staccato visual activity provokes a frenzy of feeding in the trout, which lose all table manners and snatch and grab.

Another inducement to vigorous surface capture is where particularly large, energy-rich morsels arrive on the surface. Crane flies, grasshoppers, and other largish "terrestrials" are good examples here.

QUESTION 3: Are trout slashing at the surface? If yes, continue. If no, go to question 4.

If trout are splashing around yet there is no evidence of associated winged insects it means that the food is near the surface and almost certainly fairly fast moving, yet not in the process of emerging into a winged adult. A likely candidate is small prey fish.

Trout will often herd unfortunate prey fish into the margins of pools and lakes and behave very much like sharks and other marine predators by careering through the shoal of small fish, slashing and stunning as they go. They then return to mop up the dead and disabled.

The aim must be to mimic either a fleeing fish or a dead one, belly up on the surface, see pages 184–185. Choose a

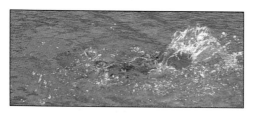

A vigorous rise. Are the trout feeding on insects at the water surface?

pattern that is of similar size and color to the actual prey fish, cast into the feeding area and retrieve the lure erratically and jerkily, all the while moving the rod tip from side to side to produce sudden changes in direction. Alternatively, choose a pattern such as a floating fry and let it lie on the surface.

QUESTION 4: Are trout "browsing" just below the surface, perhaps showing their dorsal and tail fins and sipping the surface film (i.e. head and tail rises)? If yes, continue. If no, then you have reached a dead end and probably missed something. Look again!

Such a pattern is typical of trout that are feeding on the pupae and sometimes larvae of midges, or indeed any other high concentration of slow-moving prey close to the surface. You might be unlucky and find that the fish are feeding on minute prey such as *Daphnia*; there is not much you can do, except watch in frustration or resort to lures.

Midges offer a much better prospect, however. They are members of the insect order Diptera, or true flies. The most commonly encountered species are from the family Chironomidae. Chironomid pupae (known as buzzers by anglers) may reach high densities in the surface film and sub-surface layers of rivers and lakes. The trout graze on this planktonic pasture, sucking in pupa after pupa while slowly swimming along just below the surface.

If trout are feeding in this manner, it is worth using the pond net to scoop the surface of the water near the bank, especially downwind. Look for pupae and perhaps empty skins, or "shucks." Also look out for mating swarms of adult midges, the presence of which often indicates that pupae are busy emerging nearby, and will give an indication of the size and color of buzzer to use. Go to pages 124–128.

QUESTION 5: Can you see the trout, even though they are not feeding at the surface? If yes, continue. If no, got to question 6.

If you are fishing water that is clear enough to see the trout, then sit and watch what the fish are doing. Be patient: it is far better to deduce from observation how best to entice a trout

*A Brown Trout cruising the shallow margins.
What is it eating?*

than it is to cast the first fly that comes to hand and likely scare the fish into ignoring all subsequent offerings.

Is your target fish stationed or patrolling? Can you see it feeding (watch for a glimpse of white as the trout opens its mouth)? If so, is it taking food from the bottom or from midwater? Does its behavior suggest that the food is fast moving or relatively sedentary?

A fish which moves along at a leisurely pace regularly inhaling objects either on the bottom or on weeds could be ingesting caddis larvae, mayfly nymphs, or snails. If it chases its prey over short distances it may well be taking amphipod crustaceans or the nymphs of dragonflies or damselflies. More vigorous pursuit could indicate feeding on small fish, water beetles, water bugs, tadpoles, or crayfish. Use your sweep net and pond nets to gain further clues. Turn to pages 40–43 for help with identification.

QUESTION 6: **Are there winged adult insects apparent, even though trout are not taking them from the surface? If yes, continue. If no, go to question 7.**

Use the sweep net to catch whatever is flying or sitting on the vegetation beside the water, then go to the identification key on pages 40–43. If you find adults of any of the major orders which are aquatic during their immature stages, then it is worth fishing an appropriate larval or nymphal pattern. The adult will provide some indication of the size, form, and color of the immature. As a general guide, use a version that is about the same length and overall shape as the adult minus its wings. Fish it in a manner which befits the natural insect's behavior. Try systematically varying the size and color of the imitation and also where, how and at what depth you fish it.

The larval or nymphal forms of caddisflies (Trichoptera), mayflies (Ephemeroptera), stoneflies (Plecoptera), alderflies (Megaloptera), and damselflies and dragonflies (Odonata) live on the bottom rather than in midwater or the surface, and should be presented accordingly.

In fact, not many prey organisms make their living in midwater regions. If they do end up there, they have usually either been displaced or are on their way to the top or the bottom. In rivers, food items of various sorts are dislodged from weed and gravel beds by the current and trout may well sit downstream ready to pick up hapless nymphs, larvae, and other items as they wash past.

Deciding where to fish is not so difficult as it would appear. The basic rule is simple: fish will be found where they get the best food returns for energy expended, and at least risk of being eaten themselves. Look for areas which either bear the greatest concentrations of prey or are positioned such that food will arrive on the current from upstream.

Weedbeds and rushes are sites of abundance of prey, as well as affording protection for trout. Inlets and outlets of lakes and reservoirs, and the heads and tails of pools in a river are also likely sites as they provide a passing tide of displaced prey. The downwind bank of a lake, pool, or reservoir is also an obvious spot to fish, since food animals in the surface layers of the water become concentrated there.

In short, learning to "read" a water is largely a matter of common sense and will repay the effort many times over.

QUESTION 7: **There are no winged adults evident, there is no (or only rare) indication of trout feeding at the surface, and no trout can be seen in the water because it is too murky. What on earth do I do now?**

If you have reached this point in the key, don't despair. You are now forced to rely on two types of evidence: that gained from reading the water and from sampling with the pond net. Choose the most likely spots for fish to be (see *Question 6*) and identify the common food animals living there, or at least in the regions that are accessible with the pond net.

There is not a great deal of point in fishing an imitation at or near the surface, although it might be worth trying to tempt fish to rise with a large dry fly.

If fish are not feeding near the surface, then they are either not feeding at all, or they are eating something farther down in the water.

Begin a methodical and systematic search of the bottom. For example, choose a likely spot and cast out in front. Retrieve the fly in an appropriately natural manner and recast to the same spot twice more. Next turn and cast 45° to your left and repeat the process, then 45° to your right. You will now have covered an arc of water. If you have done so without reward, vary the pattern of retrieve, then the pattern of fly. Work in this way through the various life forms which you have discovered in your pond net.

If you have still had no success, then move to a different site and start again.

Having searched the likely lies, and changed the size, color, and pattern of fly in a systematic and methodical way, yet all to no avail, then it is time to try nonimitative lures. Pursue the same strategy, but add another dimension to the search: depth. Use the "countdown" method, whereby you leave the lure to sink for different periods before retrieving it. Initially, try using retrieve rates and patterns that would suit a small fish, which might, for example, be cruising (either slowly or quickly) at a constant speed, or escaping in rapid darts. If that fails, then forget natural prey and start to use your imagination! Whenever the first trout is captured, investigate its stomach contents. Perhaps there will be nothing there but sludge, indicating that it has been feeding on very small prey such as *Daphnia*. In this case there is little chance of success with a nymph or other imitative pattern. Alternatively, the gut may contain recognizable remains, whereupon you can return to fishing a natural imitation.

If, after all that, your fish bag lies empty and forlorn, it is time to enter "blank" in the return book and seek consolation.

IDENTIFICATION KEY—
LARVAE AND NYMPHS

While you will not be able to identify everything you find, it should be possible to place all the aquatic insects encountered, whether immature or adult, in their order. Identification more accurate than this is unnecessary as far as successful fishing is concerned but anyone who finds fish prey more interesting than the fish will have no difficulty in finding suitable field guides.

We assume that fish, tadpoles, spiders, and snails will be so familiar that they can be omitted. Insects present much more of a test.

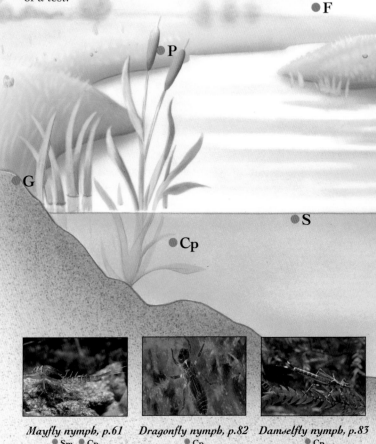

Mayfly nymph, p.61
● Sm ● Cp

Dragonfly nymph, p.82
● Cp

Damselfly nymph, p.83
● Cp

Stonefly nymph, p.91
● Cb

Alderfly larva, p.109
● Cb

Diving Beetle larva, p.116
● Cb

Midge pupa, p.123
● Su

Crane fly larva, p.136
● G

Mosquito larva, p.129
● Su

KEY

G	=	on ground
P	=	on plants
F	=	flying
S	=	on surface of water
Ss	=	swimming on surface
Su	=	swimming just under surface
Sm	=	swimming midwater
Cb	=	crawling on bottom substrate, rocks, etc.
Cp	=	crawling on submerged plants

● **Ss**   ● **Su**

● **Sm**

● **Cb**

Bloodworm, p.123
● Cb

Caddisfly larva, p.148
● Cb ● Cp

41

IDENTIFICATION KEY –
ADULTS

● F

● P

● G

● S

● Cp

Grasshopper, p.93
● P ● G

Stonefly, p.89
● P

Damselfly, p.80
● F ● P

Water Boatman, p.103
● Su ● Sm

Pondskater, p.101
● Ss

Diving Beetle, p.115
● Sm ● F ● P

Alderfly, p.107
● P

Caddisfly, p.149
● F ● P

Mayfly spinner, p.64
● F ● P

Mayfly dun, p.62
● P

Midge emerger, p.122
● F ● P ● S

KEY

G = on ground
P = on plants
F = flying
S = on surface of water
Ss = swimming on surface
Su = swimming just under surface
Sm = swimming mid-water
Cb = crawling on bottom substrate, rocks etc.
Cp = crawling on submerged plants

● Ss   ● Su

● Sm

Mosquito, p.124
● F ● P

Crane fly, p.135
● F ● P

March fly, p.140
● F ● P

INTRODUCTION TO THE
INSECT PREY

Within the animal kingdom, insects reign supreme in terms of species richness and sheer numbers. More than 50 percent of all the known organisms on Earth are insects. Many estimates point to there being many millions more insect species than we have named so far. Despite not knowing exactly how many species there are, we are sure that, without them, global ecosystems would rapidly disintegrate. On land, for instance, there would be no flowering plants, and the disruption to nutrient recycling processes and food chains would have devastating and far-reaching effects. The majority of vertebrate animal species on Earth rely on insects as a food source, some exclusively so. Although insects arose from a sea-dwelling, wormlike ancestor around 600 million years ago, very few are associated with marine habitats today. They are, however, very important in freshwater ecosystems where they form an important component of the diet of other insects, reptiles, amphibians, birds and, of course, fish. More than 2,500 years ago someone hit on the idea of catching fish by tying imitations of their insect food to a hook.

Insects are classified in 28 separate orders. Some of these, such as the dragonflies or beetles, will be well known, while others will be much less familiar.

By convention, in insect groups that have the word *fly* as part of their common name, but are not true flies (Diptera), *fly* is written together with the rest of the name as one word. For instance, butterflies, caddisflies, and stoneflies are not members of the order Diptera but crane flies, shore flies, and black flies are.

The following insect orders, which are identified by the symbol ≈, contain species which are associated, partly or wholly, with freshwater habitats during some part of their life cycle. Insects belonging to nonaquatic orders may also end up as trout food by falling on to and being trapped by the surface film.

KINGDOM: Animalia
PHYLUM: Arthropoda
CLASS: Insecta

SUBCLASS I: APTERYGOTA
These wingless insects do not undergo metamorphosis and the young stages look just like the adult. These are the most primitive of all insects and do not concern the trout angler.

Bristletails (Archaeognatha)
Silverfish (Thysanura)

SUBCLASS II: PTERYGOTA
Although the name means the winged insects, some species have lost their wings due to a parasitic lifestyle.

DIVISION I: EXOPTERYGOTA
In these orders, the young stages are called **nymphs** and look similar to the adults except for the possession of wings and sexual organs. Their wings develop on the outside of the body and metamorphosis is termed simple or incomplete. The aquatic, immature stages of mayflies, dragonflies, and stone-flies, which are often called **naiads**, do not resemble the adult stage nearly so closely as do the young of terrestrial insect groups such as the bugs or the cockroaches.
≈ Mayflies (Ephemeroptera)
≈ Dragonflies and Damselflies (Odonata)
   Cockroaches (Blattodea)
   Angel Insects (Zoraptera)
   Rock Crawlers (Grylloblattodea)
   Web Spinners (Embioptera)
   Termites (Isoptera)
   Mantids (Mantodea)
   Earwigs (Dermaptera)
≈ Stoneflies (Plecoptera)
   Grasshoppers and Crickets (Orthoptera)
   Stick Insects (Phasmatodea)
   Booklice and Barklice (Psocoptera)
   Parasitic Lice (Phthiraptera)
≈ Bugs (Hemiptera)
   Thrips (Thysanoptera)

DIVISION II: ENDOPTERYGOTA

In these orders, the young stages, which are maggot, grub, or caterpillar-like, are called **larvae** and look very different from the adults they will become. Their wings develop internally and metamorphosis is termed complete. The incredible transformation from larva to adult takes place during a pupal stage. These orders contain the most advanced of all insects.

≈ Alderflies and Snakeflies (Megaloptera)
Lacewings and Antlions (Neuroptera)
≈ Beetles (Coleoptera)
Strepsipterans (Strepsiptera)
Scorpionflies (Mecoptera)
Fleas (Siphonaptera)
≈ True Flies (Diptera)
≈ Caddisflies (Trichoptera)
Butterflies and Moths (Lepidoptera)
Sawflies, Wasps, Bees, and Ants (Hymenoptera)

An adult caddisfly resting on a twig.

HOW TO CREATE AND USE THE ARTIFICIAL FLIES

The selection of fly patterns provided have been chosen with three important criteria in mind.

First, the patterns should have been proven to catch fish. We have not been slaves to tradition or fashion: if a pattern works well it matters not a jot whether it is 200 years old, or designed last year and constructed of some modern material.

Second, the pattern should be, to some degree, imitative of the real animal. This is a difficult thing to prove, since we cannot know exactly what a trout perceives. Nevertheless, all the patterns used are known to be successful in catching trout when used under conditions where (a) the natural prey animal is present, and (b) the artificial is fished in a manner that reflects the behavior of the prey. Providing that it is of the right sort of size, shape and color, and is fished in an appropriately lifelike manner, a given fly could be used as an imitation of numerous insect species from different orders.

Third, the flies should be relatively easy to tie, or be readily available commercially. A basic recipe of the materials is provided as a guide to creating your own variations. The hook codes used refer to the Tiemco or Mustad range. Introductory fly-tying instructions follow on pages 48–53.

We also provide some advice on how to make the artificial fly behave like a real insect, perhaps exaggerating its behavior, but never making it do totally unfeasible things. You may well catch trout on occasions with a mayfly nymph stripped through the water at a rate of 1 yard per second, but it is doubtful that the trout mistook it for the real thing. The instructions we give are therefore based on the natural behavior of the prey animals. You should also use your own eyes and replicate what *you* see the animals doing.

Most of the instructions are based on the assumption that you are fishing in still water. The current adds another aspect to river and stream fishing. Not that this makes much difference to the underlying rule: don't make the fly do anything that the natural animal couldn't do.

PHEASANT TAIL NYMPH

The Pheasant Tail Nymph offers the classic nymph profile. Most often tied on hooks from size 10 down to as small as size 20, it makes a superb imitation of a range of mayfly nymph species.

The Pheasant Tail Nymph is usually weighted with turns of copper wire wound under the thorax to help it sink quickly to the target fish's depth. Variations may be created by substituting different colored dubbing materials, such as olive and orange, for the pheasant tail thorax. Also, tied in larger sizes than normal, at a pinch the Pheasant Tail Nymph can be used to suggest a stonefly nymph or a damselfly nymph.

1 Fix the hook securely in the vice and wrap with brown thread from the eye down the shank to a position opposite the

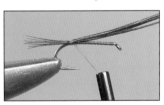

barb. Select six or seven long strands of rich chestnut-colored cock pheasant tail fibers. With the thread, tie them in at the hook bend so the tips project back as a tail.

2 Take 3 inches of fine copper wire which will form the ribbing. Tie it in at the same point as the pheasant tail fibers,

allowing the waste end of wire to lie along the length of the hook. Covering this short stub of wire secures the ribbing material and forms an even base on which to wind the body.

3 Wind the thread up to a point three quarters along the

hook shank. Take hold of the butts of pheasant tail fibers which were used for the tail and wind them, without twisting, over the hook. Let the fibers spread flat as they are wound.

4 Wind four open turns of the copper wire over the pheasant tail body. Bind the butts of pheasant tail fibers back (these

will later form the wing cases) then catch in a second 3 inch length of copper wire at their base. Wind touching turns over the hook to produce a bulbous, weighted thorax.

5 Catch in, by the tips, a second bunch of cock pheasant tail fibers at the rear of the copper wire thorax. Take the thread

up to just behind the hook eye. Then, twist the cock hackle fibers gently before winding them up to the eye. Secure the butts in place with thread and remove the excess.

6 Finally pull the remaining pheasant tail fibers over the back of the thorax securing them too, with turns of thread, just behind the eye. Trim the excess butts of pheasant tail fibers with sharp scissors before building up a small neat head with

the thread. Whip finish and add a coat of clear varnish to the head.

BLUE WINGED OLIVE –
THORAX TIE

The most widely used groups of dry flies are those that imitate the sub-imago, or dun, of the smaller mayfly species.

The Blue Winged Olive is a superb imitation of a number of similar species all of which have an olive colored body and smoky gray wings. This particular dressing is known as a "thorax" tie because the hackle is wound the length of the thorax rather than simply as a collar. It is a style of tying that keeps the hackle sparse and the overall effect more lifelike.

DRESSING

HOOK: TMC 100, sizes 14–20	**WING:** Two dark gray cul-de-canard plumes
THREAD: Olive pre-waxed	
TAIL: Four blue dun cock hackle fibers	**THORAX:** Medium olive Antron or dubbing blend
BODY: Medium olive Antron or dubbing blend	**HACKLE:** Blue dun cock hackle

---

1 Secure the hook in the vice and wrap with well-waxed olive thread from the eye to a point slightly round the bend. Dub a tiny pinch of Antron, dyed medium olive, on to the

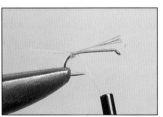

thread and wind it to form a small ball. Next, select four blue dun cock hackle fibers and catch them in in two pairs either side of the dubbing ball to give a tail with a V-profile.

2 Take a second, larger pinch of dyed medium olive Antron

and using a simple finger-and-thumb twist, dub it evenly on to the prewaxed thread. When dubbing fur always keep twisting in the same direction creating a thin, slightly tapered Antron rope. Wind the rope two thirds of the way up the hook shank to form the fly's body.

3 Take two dark gray cul-de-canard feathers. These soft

duck feathers, impregnated by the bird's natural water-proofing oils, float superbly and make an ideal wing. Lay the plumes together, with tips even, and tie in the prepared wing.

4 Select a blue dun cock hackle with fibers slightly longer than the width of the hook gap. Remove any soft or broken

fibers from the base and trim to leave a short stub of bare hackle stalk. Use two turns of thread to tie the prepared hackle in at the front of the body by this stub.

5 Dub on a third pinch of medium olive Antron, winding it

two turns behind the wing and then three turns in front to form a thorax. Next, take hold of the hackle by its tip, and wind it with a pair of hackle pliers in four open turns to a point just behind the eye.

6 Finally dub a tiny pinch of medium olive Antron on to the thread, build a small, neat head. Tie off with a whip finish.

Ensure that the wing of the finished fly lies back at approximately 45° and that the tip projects just past the hook bend.

ELK HAIR CADDIS

Hatches of caddis, or sedges, produce some of the most spectacular mid to late season trout fishing to be had. One of the simplest imitations is Al Troth's Elk Hair Caddis, which uses a wing of cow elk hair to produce a pattern which floats superbly, even in broken water, and which has the typical roof-wing profile, diagnostic of a natural caddisfly. In this original dressing the body comprises hare's fur but may be substituted by pale green, tan, or amber fur to deadly effect.

DRESSING

HOOK: TMC 100, sizes 10–18

THREAD: Brown pre-waxed

RIB: Fine gold wire

BODY: Dark hare's fur (either pale green, tan, or amber fur may be substituted)

HACKLE: Brown or Furnace cock hackle

WING: Bleached cow elk

1 Fix the hook securely in the vice and wrap with brown tying thread from the eye to a position opposite the barb. At

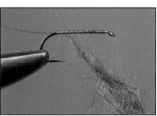

this point tie in 2 inches of fine, gold wire. Wax the thread well with sticky, fly tyers wax then take a pinch of dark fur, taken from a hare's ear, spreading it along the thread.

2 Between finger and thumb gently twist the fur on to the thread, in a technique known as dubbing. Always twist in the

same direction to build up a neat, slightly tapered rope. The sticky wax will help the fur to adhere. Wind the fur rope from the hook bend to a point just short of the eye.

3 Select a natural brown or Furnace cock hackle (a reddish-brown feather with a black center) with fibers a little longer

than the hook gap. Remove any broken or downy fibers from its base to leave a short section of bare stalk. Tie this in with tying thread just behind the eye.

4 Grasp the tip of the hackle with hackle pliers, and wind the hackle in closely spaced turns down to the hook bend. The

hackle should cover the entire body. Secure the hackle in place by winding the gold wire ribbing up through it so each turn of wire locks a turn of hackle stem.

5 Remove the excess hackle tip and gold wire with a pair of scissors before cutting a bunch of bleached elk hock fibers from the skin. Ensure that the tips of the elk hair are even

before laying them over the top of the body and holding in position with two thread turns. The tips of the hair should project just past the hook behind.

6 Secure the wing in place with tight turns of thread. The hair will flare as the thread pulls tight, but once the wing is

held fast, use two or three less tight turns of thread over the base of the wing to make it lie low over the body. Tie off with a whip finish. Trim the excess hair over the eye leaving a short stub.

PART II
THE ANGLER'S IDENTIFIER

KEY FOR SPECIES LOCATION

N.A.–North America; *U.K.*–United Kingdom; *Aus*–Australia.

MAYFLIES

THE NATURAL INSECTS

ORDER Ephemeroptera
DERIVATION Greek: *ephemeros*–lasting a day; *pteron*–a wing
SIZE Body length up to 1½ in. Wingspan up to 2 in
NUMBER OF SPECIES World—2,500; N.A.—611; U.K.—46; Aus—85

THE CULPRITS

It must be said that this group is responsible for much of the unnecessary complication in trout fishing. There are literally thousands of patterns of fly tied to imitate these insects and much of trout fishing lore and jargon has built around them. There are two important reasons why the beginner can ignore all this and keep things very simple.

Firstly, unlike an order such as the true flies (Diptera), which differs enormously between species and families in all aspects of appearance and biology (contrast, for example, the families Tipulidae (crane flies) and Culicidae (mosquitoes)), the mayflies all look and do pretty much the same. All the angler, and the trout, need to know is that what is before them is a small brown mayfly or a large yellow one. There are, of course, significant variations in size, body form, and behavior between species, seen especially in the nymphs, but these are best imitated by direct observation, without having to identify the species or even family of mayfly concerned.

Secondly, while it may be an easy matter to distinguish a crane fly from a mosquito within the order Diptera, it is just not possible for the lay person to distinguish the groups of mayflies with any certainty. For instance, to identify an adult mayfly as a member of the family Ephemerellidae requires asking 14 consecutive, technical questions, which need specialist knowledge and a microscope to answer.

What we have chosen to do is to provide, for interest, some background biological information on five of the common mayfly families. Additionally, we have given a selection of fly patterns, representing the major life stages of a typical mayfly. Keep a range of imitations in your fly box, tied from this or any other list of artificials, and use the one that looks most like the species actually present at the waterside.

Mayflies are delicate insects with cylindrical bodies, slender legs, and two pairs of richly veined wings, which are held vertically at rest. This resting posture has led to an alternative common name, the up-winged flies. They are found all over the world especially in temperate regions.

The head has short, hairlike antennae, large, compound eyes, and very weak biting mouthparts. The adults do not feed, and live for a very short time. Most adults live less than a day, and in some species they survive for only a matter of minutes. The front pair of wings are large and triangular whereas the smaller, hind wings are more rounded. The legs are elongate and the slender abdomen bears two or three long filaments.

While the adults of some mayfly species may be found at almost any time of year, others emerge only during particular

A mayfly subimago (or dun) resting
before it molts to become adult (Baetidae).

months; May to August being favoured by many species. The name mayfly applies to all members of the order, not just to those species emerging in May.

Emergences may be spectacular and highly predictable locally, or occur throughout much of the year. We have not included information on hatch times for two reasons. Firstly, emergence times are specific to particular species of mayfly, and we are not providing this level of taxonomic detail. Secondly, the same species can emerge at different times of the year according to locality, and we are aiming to provide a guide which can be used anywhere. We would rather that the angler learn to identify a mayfly as a mayfly when found, and then make the decision to fish an appropriate pattern of artificial, than to rely on what *should* be present according to a textbook.

NYMPHS

The aquatic nymphs live in streams, ponds or lakes and most prefer flowing water. Nymphs, which generally have lateral abdominal gills and three tails, vary in appearance according to habitat. Free-swimming species are elongate with long legs, while crawling species which live in faster flowing water tend to be flat and squat with short, strong legs. The nymphs of some species burrow in sand and silt. Mayfly nymphs use their mouthparts to scrape a wide range of plant and animal matter from underwater surfaces, while some filter fine food particles from the water.

The Ephemeroptera are the oldest group of winged insects on Earth today. They are unique in having a pre-adult winged stage called the subimago or dun; that is to say, they are the only insects which moult again after they have developed functional wings.

Having moulted anything from 12 to more than 50 times and taken up to two years to reach adulthood, fully grown nymphs swallow air and rise to the surface of the water where they moult into the winged, subimaginal stage. This is one of the most hazardous times for any mayfly as it is very vulnerable to attack from fish below and from dragonflies and other predators above. The mass emergences that take place at dawn or dusk reduce the chances of any single animal

*A newly emerged mayfly adult (or spinner) struggles free
from its subimaginal skin* (Ephemeridae).

being seen and eaten. The dull and slightly hairy subimago
flutters to rest on nearby vegetation.

The final moult to the hairless, shiny-winged adult form
takes place in anything from a few minutes to a couple of
days. Adults are known as spinners and their prime
motivation is to mate.

After mating, female mayflies drop egg masses into the
water. In some species females may crash land onto the
surface film releasing eggs in the process. The females of
other species may actually enter the water and swim down to
attach their eggs to submerged objects. The males of many
species take part in mass mating swarms over water at dusk.
The rising and falling flight pattern of these characteristic
displays attracts newly-emerged females. Males die shortly
after mating and females die after egg-laying.

Mayflies do not have any special protective mechanisms or
devices and are readily eaten by just about every aquatic
predator. What they lack in defence they make up for in very
large numbers. In a typical stream habitat there may be
anything from a few hundred to many thousands of nymphs
per square metre (square yard). Mayflies are thus an
extremely important component of all freshwater food chains.
Most mayfly species are intolerant of various types of
pollution such as nitrate fertilizer and salt run-off from fields
and roads and are therefore a good indication of clean water.

SMALL MAYFLIES

FAMILY Baetidae
SIZE Body length up to ¼ in. Mostly ¼ in.
NUMBER OF SPECIES World—800; N.A.—147; U.K.—14; Aus—13.

The front wings are elongate and oval with simple venation but hind wings of some small mayflies can be reduced or absent. Small mayflies may be variously colored, from light brown to black with yellowish, gray or white markings. The eyes of males are completely divided into upper and lower portions. The posterior end of the abdomen carries two very long tails.

A Small Mayfly dun (Baetidae).

These mayflies are rarely seen far from streams, rivers, ditches, ponds, and lakes. Species such as those belonging to the genus *Baetis* are a typical component of freshwater streams and rivers all over the world. Species vary in their habitat preferences, some living in warm, static or slow-moving water while others prefer cold fast-flowing waters. Some Small Mayfly species can tolerate polluted waters, unlike other mayfly families. These species can be found at higher altitudes and latitudes.

The aquatic nymphs, which are usually slender with a streamlined shape, are herbivorous and generally live among vegetation or under stones and debris. They are active swimmers and climb about on submerged plants. Adult females will enter the water, even going through waterfalls to lay their eggs on rocks. *Cloeon dipterum* gives birth to nymphs not eggs. The females will remain hidden in vegetation while their eggs develop inside their bodies. The young nymphs are dropped into the water below, where they swim down to the bottom to commence feeding.

A newly molted Small Mayfly adult (or spinner) (Baetidae).

The nymph of a Small Mayfly (Baetidae).

On the water surface a Small Mayfly dun begins to emerge from its nymphal skin (Baetidae).

BURROWING MAYFLIES

FAMILY Ephemeridae
SIZE Body length up to 1¼ in
NUMBER OF SPECIES World—150; N.A.—13; U.K.—3; Aus—0.

The wings of these large mayflies are clear or brownish in color although a few species (*Ephemera* spp.) have dark-spotted wings. The bodies of these mayflies are often pale-colored with characteristic dark markings and well-developed legs. This family contains some of the world's largest mayflies.

The aquatic nymphs, whose front legs are adapted for digging, are found burrowing in sand and silt at the bottom of streams, rivers, lakes or ponds. Some species can burrow to incredible depths of 50 ft or more. They have a shovel-like process on the head and long, toothed mandibles for moving the silt which is then pushed backwards by the legs. The nymphs eat organic material extracted from their burrows.

A female dun Burrowing Mayfly (Ephemeridae).

A mayfly spinner (top with dark, shiny wings) and dun of the same species share a perch. The adult's cast subimaginal skin remains attached to the reed below (Ephemeridae).

The mass emergences of some species such as *Hexagenia bilineata* around the banks of the Mississippi can cause a hazard for road and river traffic as the bodies of vast numbers of adults form a slippery layer several inches thick. Sadly, in some areas where these insects were once very common, pollution has reduced their numbers dramatically. Burrowing mayflies are found particularly across the Northern Hemisphere and in the African and Oriental regions.

SPINY CRAWLER MAYFLIES

FAMILY Ephemerellidae
SIZE up to ½ in
NUMBER OF SPECIES World—170; N.A.—78; U.K.—2; Aus—1.

Ephemerellids are widely distributed throughout the northern hemisphere and although associated with all manner of aquatic habitats they are especially found in running water. Adults are variable in size and color with no easily recognisable distinguishing features.

Typically, mated females lay their eggs on the water surface. Nymphs are also variable in appearance, ranging from slender to robust and from slightly to strongly flattened with, in many species, short spines on the head and thorax. They are relatively poor swimmers and can be found crawling among weed, under rocks and over the bottom. Many ephemerellids feed as nymphs on a range of fine, particulate, organic matter which they filter from the water, while others scrape algae and similar material from the substrate. Some species are even carnivorous.

An adult Spiny Crawler Mayfly (Ephemerellidae).

STREAM OR FLATHEAD MAYFLIES

FAMILY Heptageniidae
SIZE Body length up to ½ in
NUMBER OF SPECIES World—550; N.A.—133; U.K.—11; Aus—0.

The adults are mostly dark brown with clear wings. Some species can be yellow or reddish-brown with black, white, or yellow markings. In many species the wing veins are dark brown. Unlike the males of Prongill and Small Mayflies, the eyes of male Stream Mayflies are not divided into upper and lower portions. The posterior end of the abdomen carries two long tails.

These mayflies, which are distributed widely throughout the Northern Hemisphere, are usually associated with fast-flowing water such as mountain streams, although some species can be found around ponds and the margins of lakes. The dark, aquatic nymphs live under stones or logs and some-times in submerged vegetation or bottom debris. They have flattened heads and bodies with the eyes and antennae on the dorsal surface. The nymphs of most species are active but some are poor swimmers and cling to the substrate by means of a gill holdfast. Although species in some genera are carnivorous, nymphs of most species feed on algae, which they scrape from the surface of rocks, or plant material.

A Flathead Mayfly nymph (Heptageniidae).

PRONGILL MAYFLIES

FAMILY Leptophlebiidae
SIZE Body length up to ½ in. Mostly ¼ in
NUMBER OF SPECIES World—600; N.A.—70; U.K.—6; Aus—54.

The color of the adults is variable from light to dark brown. The wings have dark brown longitudinal veins in most species. The eyes of males are strongly divided, the upper area having large lenses or facets and the lower area having smaller facets.

These insects are found near slow-flowing streams and rivers or near the margins of ponds and lakes. In contrast to Stream Mayflies this family is largely distributed throughout the Southern Hemisphere. More than 70 percent of all Australian mayfly species belong to the family.

The crawling nymphs of this widely distributed and common group like to live in crevices under stones and logs or in debris. The nymphal body shape is very variable and ranges from flattened to relatively robust. The abdominal gills, which can take the form of multifingered lobes or flaps, are as variable in shape as the design of the body itself. Nearly all are herbivores, capable of shredding a variety of plant materials, wood, or detritus.

A Prongill Mayfly nymph (Leptophlebiidae).

THE ARTIFICIAL FLIES

MAYFLY NYMPH
(WALKER'S)

DRESSING

HOOK: TMC 5262, size 10-12	**RIB:** Brown thread or floss
THREAD: Brown	**THORAX:** As body with fibers
TAIL: One bunch of pheasant	picked out
tail fibers	**WINGCASE:** Pheasant tail fibers
BODY: Cream or off-white wool	with ends turned down for legs

FISHING METHOD

A highly successful pattern representing the nymphs of larger mayfly species with a more cylindrical body form. Move the fly slowly along the bottom, or ascending to the surface to emerge.

ANGLERS' NAME
BLUE-WINGED OLIVE NYMPH

DRESSING

HOOK: TMC 100, size 12	**BODY:** Blue dun wool
THREAD: Gray	**THORAX:** Light gray wool
TAIL: Blue dun hackle fibers	**WINGCASE:** Pheasant tail fibers
TAG: Red wool	**HACKLE:** Pale blue dun

FISHING METHOD

Fish slowly ½ in per second along the bottom, or in the surface film if adults are emerging. Since some baetids are active swimmers, the fly can be darted along at ¾–1½ in per second.

MARCH BROWN NYMPH

DRESSING

HOOK: TMC 100, size 12–14	**THORAX:** Hare's ear
THREAD: Brown	**WINGCASE:** Hen pheasant
TAIL: Brown partridge	**LEGS:** End pheasant fibers
BODY: Hare's ear	dressed back
RIB: Fine gold wire	

FISHING METHOD

Either fish as a nymph (move very slowly along the bottom) or an emerging dun (twitch in the surface film).

PHEASANT TAIL NYMPH

DRESSING

HOOK: TMC 100, size 10	**THORAX:** Pheasant tail fibers
THREAD: Brown	**WINGCASE:** Pheasant tail fibers
TAIL: Pheasant tail fibers	**LEGS:** Wingcase dressed over
BODY: Pheasant tail fibers	head and down to form legs
RIB: Copper wire	

FISHING METHOD

This is a classic nymph pattern which can be tied in many sizes. It is best fished very slowly (no more than ½–¾ in per

second) to mimic a nymph walking along the bottom, pausing occasionally to feed on algae and detritus. It could also be used as an imitation of a nymph rising to the surface just prior to emergence.

ANGLERS' NAME
GOLD-RIBBED HARE'S EAR

DRESSING

HOOK: TMC 100, size 12

THREAD: Brown

TAIL: Hare's mask (long guard hairs)

BODY: Dubbed hare's ear

RIB: Flat fine gold tinsel

FISHING METHOD

This pattern could be used either as a nymph or as an emerging dun. Hence, either fish very slowly along the bottom or near the surface.

ANGLERS' NAME
MAYFLY EMERGER

DRESSING

HOOK: TMC 5262, size 10–12

THREAD: Brown

TAIL: Three tips of white ostrich

BODY: Cream dubbing blend

RIB: Brown thread or floss

THORAX: Cream dubbing blend

WINGS: A few natural or white cul-de-canard feathers tied in by the butts and pulled down to proportion

FISHING METHOD

A marvelous imitation of an emerging mayfly. Fish the pattern so that the body is just below the surface and the wings project above. Imagine that the insect has only just

struggled free from its nymphal shuck and is in the process of inflating and expanding its wings. This is a particularly vulnerable time for a mayfly. Leave the pattern to drift.

OLIVE COMPARADUN

DRESSING

HOOK: TMC 100, sizes 14–20

THREAD: Olive

TAIL: Deer hair

BODY: Medium olive dubbing fur

THORAX: Medium olive dubbing fur

WING: Deer hair flared 180° around the top of the hook

FISHING METHOD

This pattern can be used to mimic a newly emerged dun or one in the process of struggling free of its shuck. Treat with

floatant and don't worry if it sinks a bit. The wings provide a useful surface indicator. Leave the fly to drift or twitch it to represent the emerging insect's struggles.

MAYFLY DUN

DRESSING

HOOK: TMC 5262, size 10–12

THREAD: Brown

TAIL: Three pheasant tail fibers tied long

BODY: Buff goose feather fibers

RIB: Brown thread or floss

HACKLES: 1st—dyed green partridge or mallard breast feather; 2nd—natural red game hackle; 3rd—orange short-fiberd cock hackle

FISHING METHOD

A useful imitation of a newly emerged mayfly dun.

ANGLERS' NAME
ADAMS

DRESSING

HOOK: TMC 100, sizes 10–20

THREAD: Black

TAIL: Brown and grizzly cock hackle fibers

BODY: Gray fur

WING: Two grizzly hackle points tied upright

HACKLE: Brown and grizzly cock hackles

FISHING METHOD

Use to represent a dun or spinner of many mayfly species. After struggling free from the nymphal exuvium (shuck) the insect sits on the water surface until its wings are dry enough to take flight. Treat the fly with floatant and leave it drift on the surface. After a while pull it across the surface for a distance of about a yard to make it appear to accelerate prior to attempting to fly off. Then stop the pull abruptly, as if the insect had aborted takeoff. Leave the fly to sit for a minute or so longer and move it again.

CDC DUN

DRESSING

HOOK: TMC 100, sizes 14–20	**THORAX:** Gray, cream, or olive wool
THREAD: Gray	**WING:** Natural gray cul-de-canard
BODY: Gray, cream, or olive wool	**HACKLE:** Natural gray cul-de-canard

FISHING METHOD

Treat with floatant and fish on top of the water to represent a newly emerged dun.

ANGLERS' NAME
MARCH BROWN DUN

DRESSING

HOOK: 79704 BR, size 12–20	**RIB:** Gold wire
THREAD: Black	**WINGS:** Dark hen pheasant for male and light for female
TAIL: Partridge hackle fibers	**HACKLE:** Dark partridge
BODY: Hare's ear	

FISHING METHOD

A well-tried pattern which can be used to represent the adult (dun especially) of any small to medium-sized mayfly species.

ANGLERS' NAME
LUNN'S PARTICULAR

DRESSING

HOOK: 79764 BR, size 16–20 | **WINGS:** Two blue dun hackle
THREAD: Brown | tips tied upright and separated
TAIL: Red game fibers | **HACKLE:** Red game cock
BODY: Stripped peacock quill

FISHING METHOD

This pattern could represent either a small mayfly dun or spinner. Treat with floatant and fish on the surface. Leave the fly to drift with the current. Occasionally scoot it gently over short distances (2–4 in).

ANGLERS' NAME
BLUE-WINGED OLIVE THORAX

DRESSING

HOOK: TMC 100, sizes 14–20 | **THORAX:** Medium olive dubbing
THREAD: Olive | blend
TAIL: Blue dun hackle fibers | **WING:** Natural cul-de-canard or
tied forked | gray turkey flats
BODY: Medium olive dubbing | **HACKLE:** Blue dun cock
blend

FISHING METHOD

Fish either on the surface or in the surface film to mimic anything from an emerging dun to a dead dun or spinner (therefore leave still, twitch, or scoot occasionally, as appropriate).

MAYFLY SPINNER

DRESSING

HOOK: TMC 5262, size 10–12	**BODY:** Cream or off-white wool
THREAD: Brown	**RIB:** Brown thread or floss
TAIL: Three pheasant tail fibers tied long	**HACKLE:** One long-fibered badger cock hackle

FISHING METHOD

This is a simple but effective imitation of the adult spinner of a large mayfly species. The hackle serves as both wings and legs. Treat with floatant and fish on the surface. Let it drift with the current and pull it along for distances of a yard or so every now and then, either at a constant rate of about 1½ in per second (as the insect might do as she flutters along laying eggs) or else to mimic an insect accelerating up to takeoff velocity.

ANGLERS' NAME
GREENWELL'S GLORY

DRESSING

HOOK: 79704 BR, size 12
THREAD: Olive floss or black-waxed pale primrose thread
BODY: Olive floss or black-waxed pale primrose thread

RIB: Fine gold wire or fine gold oval
WINGS: Two pale starling wing slips tied forward and separated
HACKLE: Greenwell's hackle

FISHING METHOD

This pattern could be used to represent either a dun or a spinner. Treat with floatant and fish on the surface, leaving it to drift or occasionally giving it a gentle tug causing it to scoot a short distance.

ANGLERS' NAME
WHITE-WINGED PHEASANT TAIL

DRESSING

HOOK: TMC 100, size 14–16
THREAD: Brown
TAIL: Tips of the body material
BODY: Cock pheasant tail fibers

over a wet varnished shank
WINGS: White cock hackle tips tied vertically
HACKLE: Ginger cock

FISHING METHOD

An elegant little pattern mimicking any small dun or spinner. It should be treated with floatant and presented on the surface, leaving it to drift with the current and perhaps scooting it along for short distances of 4–6 in on occasions.

PALE MORNING DUN

DRESSING

HOOK: TMC 100, sizes 14–20

THREAD: Pale yellow

TAIL: Pale blue dun hackle fibers tied forked

BODY: Pale yellow dubbing fur

THORAX: Pale yellow dubbing fur

WING: Gray turkey flats

HACKLE: Light blue dun cock hackle

FISHING METHOD
Fish on or just under the surface and move, if at all, only short distances.

ANGLERS' NAME
PALE WATERY

DRESSING

HOOK: 79704 BR, size 14–20

THREAD: Claret

TAIL: Red game fibers

BODY: Pale watery green goose herl

WINGS: Two pale starling slips tied upright and separated

HACKLE: Red game cock

FISHING METHOD
Treat with floatant and fish on the surface, perhaps tugging it over short distances every now and then to mimic takeoff.

ANGLERS' NAME
Yellow Humpy

DRESSING

HOOK: TMC 5262, size 12	vertically then separated
THREAD: Yellow	**WINGCASE:** Deer hair tied
TAIL: See wingcase	down and dressed down as tail
BODY: Yellow floss	**HACKLE:** Red game cock hackle
WINGS: Clipped deer hair tied	with light grizzle hackle

FISHING METHOD

This pattern is a useful all-purpose dry fly pattern for large mayflies and other species. Treat with floatant and fish on the surface, letting it drift and perhaps every now and then

pulling it for distances of up to 1 yard, so that the fly accelerates across the water before stopping. This movement could represent either egg-laying or attempting to take off.

ANGLERS' NAME
CDC Spent

DRESSING

HOOK: TMC 100, sizes 12–20	**BODY:** Rusty brown Antron
THREAD: Orange	**THORAX:** Rusty brown Antron
TAIL: Blue dun hackle fibers tied	**WING:** Natural cul-de-canard
forked	tied spent

FISHING METHOD

This raggedy-looking pattern will serve as an emerging, or spent adult mayfly. Accordingly, fish on or just under the surface, imparting movement as appropriate.

ROYAL WULFF

DRESSING

HOOK: TMC 5262, size 8–12

THREAD: Black, brown, or dark gray

TAIL: Brown deer hair

BODY: Peacock herl with a band of red floss

WINGS: Two white deer hair bunches tied upright and separated

HACKLE: Chocolate brown cock

FISHING METHOD

This pattern can be tied in a range of colors and is an excellent all-purpose dry fly for representing many large-winged insects, including mayflies. Treat with floatant and fish on the

surface. Perhaps occasionally accelerate it along the surface for distances of a yard or more, or else twitch it to represent a struggling terrestrial trapped in the surface film.

SPENT MAYFLY

DRESSING

HOOK: TMC 5262, size 10–12

THREAD: Brown

TAIL: Three pheasant tail fibers

BODY: Cream or off-white wool

RIB: Brown thread or floss

WINGS: Two bunches blue dun cock hackle fibers tied sloping forward and separated

HACKLE: Badger cock

FISHING METHOD

A lovely pattern for a dead adult mayfly with its wings flattened in the surface film. Fish stationary, either on the surface or in the surface film.

ANGLERS' NAME
SHERRY SPINNER

DRESSING

HOOK: 79704 BR, size 14

THREAD: Orange

TAIL: Red game hackle

BODY: Orange floss

RIB: Gold wire

WINGS: Pale starling tied spent

HACKLE: Red game cock

FISHING METHOD

This pattern provides an excellent imitation of a small, dead adult (dun or spinner, despite its name) and should be fished stationary, either on the surface or in the surface film.

ANGLERS' NAME
BUNSE GREEN DRAKE NATURAL DUN

DRESSING

HOOK: TMC 100, size 12

THREAD: Yellow

TAIL: Two mink tail nutria or beaver guard hairs

BODY: Green foam

RIB: Natural dun deer hair

WINGS: Pale starling tied spent

HACKLE: Red game cock

FISHING METHOD

A useful imitation of a dead (spent) adult mayfly. Fish on the surface or in the surface film, leaving it to drift.

DRAGONFLIES AND DAMSELFLIES

THE NATURAL INSECTS

ORDER Odonata
DERIVATION Greek: *Odontos*–toothed (of the mandibles)
SIZE Body length up to 6 in. Wingspan up to 8 in
NUMBER OF SPECIES World—5,500; N.A.—500; U.K.—45; Aus—300

Anyone who has gone anywhere near a body of water could not fail to notice these beautifully colored insects as they wheel and dart through the air. Their large size and sometimes fierce appearance has given rise to common names such as devil's darning needles, horse stingers and snake doctors. Despite the beauty of the adults, it is the nymphs that are of primary interest to the angler.

The order is split into two major groups, the dragonflies and the damselflies, but many people use the name dragonfly in a broad sense to refer to any species in the order. These insects are particularly species-rich in tropical and sub-tropical regions. In general, the head, which is large and very

A male damselfly perching.

mobile, has biting mouthparts, short hairlike antennae, and very large compound eyes. The eyes are very sensitive to movement and give vision in all directions. In contrast to mayflies, dragonflies and damselflies are voracious predators and feed throughout their adult lives. In dragonflies, the head is rounded and the very large eyes are not widely separated, whereas in damselflies, the head can be broader than long and the eyes are widely separated. The bulky thorax is packed with the flight muscles that drive the two pairs of equally sized, long, narrow and richly veined wings. A good way of distinguishing members of the two groups in the field is that, in dragonflies,

A dragonfly basking in the sun.

the wings are held out sideways from the body at rest whereas in damselflies, the wings are held together over the abdomen. A further distinction is that the hind wings of dragonflies are large and rounded at their base whereas in damselflies both hind and front wings are of similar shape and narrow at their

In flight, the wings of dragonflies beat out of phase.

bases. Damselflies and dragonflies have three pairs of spiny legs which both act as a prey-capturing net and allow in-flight feeding. The superb flying skills of dragonflies include the ability to turn within their own body length, hovering, backward flight and speeds of up to 30 mph. The characteristic whirring sound of dragonflies in flight is due to the fact that the two pairs of wings do not beat together; instead they beat out of phase, skimming past each other at each stroke.

Mating and courtship in dragonflies and damselflies is complex. Males can be territorial and both sexes may mate several times per day. Dragonflies tend to lay their eggs on or at the water's surface or on the surface of submerged plants. They do this by dropping egg masses or strings or dipping the end of the abdomen just below the surface. Some species may enter the water to lay eggs in a stream bed. Damselflies, on the other hand, lay their eggs inside the stems and tissues of aquatic plants.

NYMPHS

The aquatic nymphs are very similar in some regards to those of the Ephemeroptera but instead of being mainly herbivorous they are all highly predacious. The mouthparts are specially modified for rapid and efficient prey capture. The prehensile facial mask, bearing the jaws, is hinged and, with a rapid

Dragonfly nymphs have a squat, robust shape.

Damselfly nymphs are slender and have terminal gills.

increase in blood pressure, can be shot forward from under the head to seize prey. Nymphs eat small insect larvae and other aquatic invertebrates and can stalk and kill larger prey items as they grow in size and power. All are camouflaged and algal growths and silt particles make the coloring even more effective. The mature nymphs of larger species can easily overcome tadpoles and small fish.

Dragonfly nymphs are fierce predators and will even catch and eat small fish.

Damselfly nymphs obtain their oxygen from the water by means of three, feathery or flap-like, external gills located at the end of the abdomen. The gills also act as paddles providing propulsion as the abdomen undulates from side to side rather like a small fish.

In dragonfly nymphs, the gills are internal, located within a rectal chamber. Water is drawn in and out to provide a respiratory current over the gills. If required, water can be forced out of the rectal chamber at high speed to provide a jet-propelled escape.

Depending on the species, conditions, and latitude, development may take anything from a few months to two or more years and the nymphs can have up to 15 molts. When fully grown, nymphs swim to or crawl up plant stems or on to rocks out of the water where they molt to become adult. The freshly emerged adult will rest near the empty nymphal skin to inflate and harden its wings before taking to the air. In the Northern Hemisphere species of dragonfly and damselfly can be seen on the wing anytime from early May to late October.

Apart from contributing to the diet of freshwater fish dragonfly nymphs are very useful in that they eat the larvae of midges, mosquitoes, and black flies.

Damselfly front and hind wings are nearly the same size and shape.

THE ARTIFICIAL FLIES

ANGLERS' NAME
DAMSELFLY NYMPH

DRESSING

HOOK: TMC 5262, size 10–12	**RIB:** Fine gold oval
THREAD: Olive	**THORAX:** Olive ostrich herl
TAIL: Olive marabou	**WINGCASE:** Olive goose
BODY: Olive goose	**EYES:** ½ in brass chain links

FISHING METHOD

Fish the pattern very slowly along the bottom, to represent the insect stalking prey, but then occasionally move it along for 4–20 in at a faster speed (up to 2 in per second). As the fly moves, its marabou tail will wiggle enticingly, just as the real insect's abdomen would do as it swims along. At times when adults are emerging, the nymphs that are ready to do so will swim up to the surface where they climb out onto reeds and sticks to shed their exuvium (shuck). It then pays to fish the nymph on a floating line, letting it sink to the bottom before retrieving smoothly at about 2 in per second so that the fly lifts up in the water toward the surface. Unweighted versions may also be swum along just below the surface.

GREEN MOUNTAIN DAMSEL

DRESSING

HOOK: TMC 300, size 10

THREAD: Olive

TAIL: Three goose biots dyed olive

BODY: Green Antron

RIB: Fine gold wire

WINGCASE: Olive marabou fibers tied upright

FISHING METHOD

Another good damselfly nymph mimic, which has plenty of movement in the marabou wingcases. Fish deep or swimming toward or along the surface.

CACTUS DAMSEL

DRESSING

HOOK: TMC 300, size 10–12

THREAD: Olive

TAIL: Olive marabou

BODY: Cactus olive chenille

HEAD: ½ in gold bead

FISHING METHOD

A deadly modern pattern which sparkles and undulates when retrieved. Use as other damselfly nymph imitations.

ANGLERS' NAME
Damsel Wiggle Nymph

DRESSING

HOOK 1: TMC 100, size 8–10	**HOOK 2:** TMC 100, size 10;
THREAD: Tan or olive waxed	weighted with lead wire and
nylon	linked to hook 1
TAIL: Olive or golden brown	**THORAX:** Brown dubbing blend
marabou fibers	**WINGCASE:** Brown turkey dyed
BODY: Brown or olive dubbing	olive
RIB: Gold oval tinsel	**HACKLE:** Partridge dyed olive

FISHING METHOD

Another solution to imitating the swimming action of a damselfly nymph is provided by this pattern. Here the abdomen is a separate hook linked in tandem by wire to the anterior hook which represents the head and thorax. Fish as for other damsel nymphs, either along the bottom or swimming up to the surface as if about to emerge as an adult.

THE DRAGON

DRESSING

HOOK: TMC 5262, size 12	**BODY:** Olive green dubbing blend
THREAD: Olive	**RIB:** Olive floss
TAIL: Three goose biots dyed olive	**HACKLE:** Brown partridge
	HEAD: Peacock herl

FISHING METHOD

The nymphs of these insects swim by ejecting water from their rectum. Accordingly, move the fly slowly along the  bottom, but then at intervals tug the line to produce a short 4 inch or so explosive burst of movement, as would occur were the insect escaping from a trout.

BLUE DAMSELFLY ADULT

DRESSING

HOOK: TMC 100, size 10	hackles tied spent
THREAD: Black	**HACKLE:** Blue dun
BODY: Floating fly line ribbed with black silk and varnished	**EYES:** Plastazote balls wrapped in white tights tied in figure-eights
WINGS: Two pairs black cock	

FISHING METHOD

A lifelike imitation of an adult blue damselfly. Use when adults are flying, and fish floating on the surface to imitate either a dead (spent) insect, or one on its last legs (twitching a bit).

STONEFLIES

THE NATURAL INSECTS

ORDER Plecoptera
DERIVATION Greek: *plectos*–pleated; *pteron*–-a wing
SIZE Body length up to 2 in
NUMBER OF SPECIES World—2,000; N.A.—470; U.K.—34; Aus—200.

Stoneflies, which are common in the world's temperate and cooler regions, are mostly drab brown or gray, slender insects with slightly flattened bodies. They are very weak fliers and seldom travel far from water. Adult stoneflies spend a great deal of time resting on rocks and vegetation by the sides of streams, rivers, and lakes and may, according to species, be active during day or night. Many stoneflies are attracted to lights after dark.

The head has weak or nonfunctional biting mouthparts, somewhat protruding compound eyes, and long, threadlike antennae.

Adult stoneflies can be found resting on rocks or streamside plants.

89

The front wings are much narrower and slightly longer than the hind wings, which are folded in pleats beneath. Stoneflies were the first insects to evolve special hinging in the wing bases that, unlike the case with mayflies and dragonflies, allows them to fold their wings back along the body. On account of tight wing-folding some species are called needle-winged or rolled-wing stoneflies. Some species have very short wings or none at all. The legs of adult stoneflies are quite broad, strong and seem widely separated.

Most adult stoneflies do not feed but some scrape algae and lichen from rocks.

The abdomen, which is covered by the wings at rest, is cylindrical and slightly flattened with a pair of short or long sensory tails, called cerci, at its posterior end.

The life span of an adult stonefly is usually less than a couple of weeks but can be as short as a few days or as long as three months. While most do not feed at all, the adults of some species feed by scraping fragments of algae and lichen from rocks.

Courtship and mating take place on the ground or on plants. Females lay slimy masses containing many hundreds or thousands of eggs into the water either by dipping their abdomens under the water or by dropping the eggs in flight. Stonefly eggs are round and sticky or spindle-

An adult stonefly resting on a rock in the stream.

shaped with special projections to ensure that they adhere to submerged objects or get stuck in crevices.

NYMPHS

The aquatic nymphs get oxygen from the water by diffusion into the body or by means of external, feathery gill tufts. Gill tufts may be present on the head, neck, thorax, the first few segments of the abdomen and the anus.

Many species prefer cold water (which holds more oxygen than warm water) and cannot survive in polluted water. As young stoneflies grow they become quite broad and flattened with noticeably developing wing pads and a pair of long tails at the end of the abdomen. Depending on the species, the nymphs eat decaying organic matter, algae, bacteria, debris, plant material, or small invertebrate animals. Stonefly nymphs do not swim; a fact that should be kept in mind when fishing imitations.

Pale, newly molted stonefly nymphs are particularly attractive to trout.

Development from the egg to the adult may take up to four years and involve anything up to 30 molts. Pale, soft nymphs that have just molted their skin and have not yet become dark and hardened are particularly attractive to fish and thus very useful as models for artificial flies.

The nymphs, which are a very important component of the diet of trout, can be mostly found in cold, clean, oxygen-rich, upland streams. Many species become adult, mate, and die from late autumn to early spring, some even during the coldest months of winter.

Stonefly nymphs get oxygen by means of feathery gill tufts on various parts of the body.

Some species of giant stoneflies in North America, belonging to the family Pteronarycidae, have acquired the common names salmonfly and troutfly, reflecting their important role as fish food and as lures for anglers chasing the fish that eat them. Trout gorge themselves when these insects emerge to mate during the early summer.

THE ARTIFICIAL FLIES

STONEFLY NYMPH

DRESSING

HOOK: TMC 100, size 12

THREAD: Black

TAIL: Two fibers besom

BODY: Stripped peacock herl

THORAX: Peacock herl

WINGCASE: Dark grouse herl

HACKLE: Grouse breast feather

FISHING METHOD

This pattern is a particularly close imitation of a typical stonefly nymph, with its two "tails" being especially diagnostic, to our eyes if not the trouts'. Stonefly nymphs of most species creep along the bottom feeding on algae and detritus. They do not swim and never move quickly. As a result, the nymph should be fished along the bottom, moving at no more than ½ in per second. Pause occasionally as if the imitation had stopped to feed. When adults are emerging the same pattern can either be fished in the surface film or (when using a floating line) allowed to sink and then made to rise toward the surface by retrieving smoothly.

ANGLERS' NAME
Yellow Sally
(WET)

DRESSING

HOOK: TMC 100, size 14–16

THREAD: Primrose

BODY: Dubbed pale yellow wool

HACKLE: Yellow cock hackle

FISHING METHOD

A wet version of Yellow Sally which can be used either as a drowned or emerging adult. Fish near the surface and either leave to drift (spent) or twitch to indicate an adult struggling to emerge from its nymphal exuvium (shuck).

ANGLERS' NAME
Willow Fly

DRESSING

HOOK: TMC 100, size 14–16

THREAD: Orange

BODY: Bronze peacock herl

WINGS: Two small medium grizzle hackles tied flat over back

HACKLE: Brown dun cock

FISHING METHOD

Fish this adult stonefly pattern on the surface, leaving it to float along with the current and perhaps occasionally moving it at a gentle rate over short distances or twitching it to indicate attempted takeoff.

STIMULATOR

DRESSING

HOOK: TMC 5263, sizes 6–14

THREAD: Fluorescent orange

TAIL Natural elk hair

BODY: Dubbed yellow fur overwound with a brown or Furnace cock hackle

WING: Elk hair

THORAX: Amber Antron

HACKLE: Grizzly cock hackle wound as a collar

Stimulator

Stonefly Adult

STONEFLY ADULT

DRESSING

HOOK: TMC 5262, size 10

THREAD: Black

BODY: Dark dyed olive hare's mask fur

RIB: Yellow silk

WINGS: Dark hen pheasant tail fibers rolled and tied down close to back and beyond the hook bend

HACKLE: Dyed olive grizzly cock

FISHING METHOD

These two patterns can be used to mimic larger species of stonefly. Treat with floatant and fish on the surface. Apart from allowing the fly to drift, it is worth attempting to imitate egg-laying and taking off by moving the fly across the surface for distances of up to a yard. Take into account that stonefly adults are not particularly elegant or fast in their movements, so attempted takeoff will be slow and clumsy.

GRASSHOPPERS
& CRICKETS

THE NATURAL INSECTS

ORDER Orthoptera
DERIVATION Greek: *orthos*—straight; *pteron*—a wing
SIZE Body length up to 6 in
NUMBER OF SPECIES World—19,000; N.A.—1,080; U.K.—30; Aus—2,900

This very large order is divided into two suborders, the long-horned grasshoppers (Ensifera), comprising, among others, the crickets and katydids, and the short-horned grasshoppers (Caelifera), comprising grasshoppers and locusts. Short-horned grasshoppers are much more common in temperate regions than the long-horned grasshoppers, which are predominantly tropical and subtropical. Orthopteran species occupy just about every possible terrestrial habitat and although, at first sight, it might not seem that they have anything to do with water, quite a few species live very close to, or in association with, freshwater habitats. When finding themselves in water, by accident or design, all Orthoptera can

A typical short-horned grasshopper.

95

Female crickets have conspicuous ovipositors.

swim very efficiently and are readily taken by trout. Orthopterans are especially common in warmer regions.

Grasshoppers and crickets may be herbivorous, omnivorous, or partly or wholly predacious. All species have biting, chewing mouthparts, an enlarged, saddle or shield-shaped pronotum (back of the first thoracic segment), toughened, narrow front wings (tegmina) with larger fan-folded hind wings (when present), and large hind legs modified for jumping. The singing ability of the males of many species is well known and the most commonly heard sounds are concerned with courtship.

Mainly nocturnal and solitary, ensiferans have thin, threadlike antennae which may be as long or longer than the body. The front wings very often look like leaves and are cryptically colored (camouflaged). The ovipositor is always prominent and sword-, sickle-, or stiletto-shaped and eggs are often laid singly or in small batches in plant tissues. Some species of katydid and cricket, including the mole crickets, can be found beside ponds and streams.

Members of the Caelifera, on the other hand, tend to be active during the day and many species tend to be gregarious. Caeliferans differ from ensiferans in that their antennae are not threadlike and never very long, the front wings do not look like leaves, and the females ovipositor is short and robust. The eggs are normally laid in the soil, protected by an egg pod. Many short-horned grasshoppers are warningly colored and can exude defensive secretions if attacked. Short-horned grasshoppers such as pygmy locusts and especially pygmy mole crickets are found in damp sandy areas by river banks and streamsides. Several species of pygmy mole cricket have special padlike expansions of the hind legs which enable them to swim across the water's surface.

Long-horned grasshoppers or crickets are often leaf-like and have thread-like antennae as shown by this katydid.

THE ARTIFICIAL FLIES

ANGLERS' NAME
LETORT CRICKET

DRESSING

HOOK: TMC 100, size 12

THREAD: Black

BODY: Black chenille

WINGS: Black goose tied flat over body

HEAD: Cropped black deer

FISHING METHOD

Crickets of most species will only end up on water by accident. However, like grasshoppers, they can swim by kicking their hind legs in unison, using the same pattern of movement as they employ when jumping. This is simply mimicked by fishing the fly on the surface and retrieving it with short pulls of about ¾ in, such that the fly moves along at an overall rate of about 1 in per second. Continue for a yard or so at a time, then let the fly drift for 10–20 seconds before recommencing the retrieve.

ANGLERS' NAME
GRASSHOPPER

DRESSING

HOOK: TMC 300, size 8

THREAD: Brown

BODY: Green polyethylene foam or Plastazote below, with brown raffine or pheasant tail fibers tied over the top to represent wings

LEGS: Two swan primary feather fibers dyed medium-brown

HACKLE: Beard hackle of 8–10 pheasant tail fibers tied short

FISHING METHOD

A lifelike imitation of a field grasshopper, which should be fished as other Orthoptera patterns by retrieving across the surface with short tugs to indicate kick-swimming.

ANGLERS' NAME
DAVE'S HOPPER

DRESSING

HOOK: TMC 5263, sizes 8–12

THREAD: Brown

TAIL: Red calf tail

BODY: Yellow polypropylene yarn with a brown cock hackle wound over and clipped short

LEGS: Two slips of brown turkey fibers each knotted once

WING: Mottled brown turkey

HEAD: Deer hair spun and clipped to shape

FISHING METHOD

A very effective pattern which can be used to imitate grass-hoppers. Employ the kick-swimming retrieve pattern described for the Letort cricket.

GREEN GRASSHOPPER

DRESSING

HOOK: TMC 300, size 10

THREAD: Fluorescent green

BODY: Insect green dubbing blend

RIB: Fine gold wire

HACKLE: Red game saddle wound palmer-style

FISHING METHOD

This pattern is not especially imitative of a grasshopper but will suffice when green grasshoppers or katydids abound among the bankside vegetation. Fish dry and retrieve in short pulls to imitate a grasshopper which is swimming along by kicking its hindlegs, sometimes stopping to rest between bouts of swimming.

BUGS

THE NATURAL INSECTS

ORDER Hemiptera
DERIVATION Greek: *hemi*–half; *pteron*–a wing
SIZE Body length up to 4 in. Mostly under 2 in
NUMBER OF SPECIES World—82,000; N.A.—9,946; U.K.—1,700;
Aus—5,700.

Bugs have elongate, needle-like mouthparts which are used to suck liquid food. Many bugs are largely herbivorous although some suck blood or animal juices.

Although many species of bug may fall onto the water's surface and may be eaten by fish, only a few bug families are aquatic and only two of these, the water boatmen and the backswimmers, are considered here as a major component in the diet of freshwater fish. Another bug family, the pond skaters (Gerridae), although common on the surface of water,

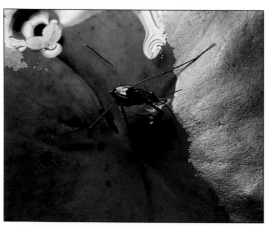

*Water-repellent hairs on the splayed middle
and hind legs of a pond skater enable
it to walk on the water surface
(Gerridae).*

are rarely taken by fish probably because they do not taste very nice.

Of the nonaquatic species, aphids and cicadas are sometimes used as models for fishing flies. Aphids are extremely abundant in all terrestrial habitats and large numbers get trapped by the surface film. The mass emergences of some cicadas at certain times may also contribute to trout diets.

BACKSWIMMERS

FAMILY Notonectidae
SIZE Body length up to ¾ in
NUMBER OF SPECIES World—300; N.A.—35; U.K—4; Aus—40

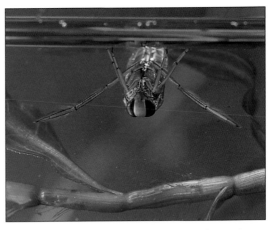

Notonectid bugs swim on their backs and can often be seen renewing their air supply at the surface (Notonectidae).

The back of these wedge-shaped bugs is pale-colored with a keel running down the middle. The front two pairs of legs are short and used for grasping food while the long hind legs act as oars.

Backswimmers swim upside down in still water in small pools, streams and around the margins of lakes and eat aquatic insects, tadpoles and even small fish. To obtain air backswimmers hang from the surface by the end of their abdomen.

WATER BOATMEN

FAMILY Corixidae
SIZE Body length about ½ in
NUMBER OF SPECIES World—550; N.A.—132; U.K.—34; Aus—31

These streamlined bugs are very similar to backswimmers but rest horizontally under the surface and swim the right way up. To conceal them better from predators the underside is pale-colored. The front legs are used for feeding, the middle legs are used for holding vegetation and the back legs provide propulsion.

Corixids can be found in the still and slow-flowing water of ponds, lakes, and streams. Water boatmen can feed on fine particulate food such as algae and plant debris as well as prey on small insect larvae. They swim rapidly but spend a lot of time holding on to vegetation. To breathe underwater they store air bubbles between their wings and the dorsal surface of the abdomen. This gives them a characteristic silvery appearance central to their imitation by artificial flies. Analyses of trout stomach contents show that these bugs are a very important dietary component.

Water boatmen swim the right way up (Corixidae).

THE ARTIFICIAL FLIES

DRESSING

HOOK: TMC 100, size 10–14	dubbing; overbody—pale orange
THREAD: Brown	raffia stretched over back and
TAG: Silver tinsel or mylar	streaked with brown
BODY: Underbody—lead foil	**RIB:** Fine silver wire
covered with dirty-white or pale	**HACKLE:** Two bunches of brown
lemon yellow angora wool	hen fibers

FISHING METHOD

Corixids (water boatmen) spend most of their time either swimming along the bottom and in among submerged vegetation, or else swimming up to and down from the surface to collect bubbles of air. Both activities can be imitated with this weighted pattern. Fish it along the bottom, retrieving it with short tugs of about ¾ in at a rate of up to 2 in per second, so as to indicate the insect swimming along with its oarlike hind legs. If using a floating line the retrieve will cause the sunken fly to lift in the water toward the surface. Once it gets there, let it sink back down (having collected its bubble of air).

LARGE BROWN CORIXA AND PLASTAZOTE CORIXA

DRESSING

HOOK: TMC 100, size 10–12
THREAD: Black
TAG: Silver tinsel
BODY: White floss or Plastazote
RIB: Flat silver wire

WINGCASE: Pheasant tail fibers (bad side)
HEAD: Black thread
LEGS: Two goose biots

FISHING METHOD

These patterns are buoyant (especially the Plastazote) and are especially good for representing water boatmen swimming along the surface collecting air. When using a sinking line it is possible to imitate a corixid diving back to the bottom with its air bubble, and then, by ceasing the retrieve, of the insect back up again when its bubble runs out.

NOTE

Patterns for corixids can be modified to imitate backswimmers by simply tying the fly upside down so that the wingcase lies under the bend of the hook rather than on top.

ANGLERS' NAME
ZINN'S CICADA

DRESSING

HOOK: TMC 5262, size 10

THREAD: Brown

BODY: Dun deer hair

THORAX: Dun deer hair

WINGS: Four blue dun hackle tips

FISHING METHOD

Adult cicadas can be extremely abundant locally at the times when the nymphs emerge from below ground where they have spent their lives (many years in some species) feeding on plant root sap. Flying adults fall onto the water and cannot swim. Fish the pattern on the surface, twitching it to indicate a struggling, drowning insect. Vary the color of the tying to represent the species found.

ALDERFLIES, DOBSONFLIES, & FISHFLIES

THE NATURAL INSECTS

ORDER Megaloptera
DERIVATION Greek: *megalo*–large, great; *pteron*–a wing
SIZE Body length up to 4 in
NUMBER OF SPECIES World—200; N.A.—64; U.K.—6; Aus—26

Megalopterans have conspicuous eyes, threadlike antennae, and biting mouthparts. The short-lived adults are commonly seen at dusk and dawn. At rest the two pairs of wings, are held in a tentlike manner over the body. There are just two families, the alderflies (Sialidae), the dobsonflies and fishflies (Corydalidae).

Adult alderflies spend a lot of time resting on waterside vegetation (Sialidae).

The aquatic larvae are slightly flattened with abdominal gill filaments. They are active swimmers and eat a range of arthropods, worms and even small fish. Larval development can take more than one year and pupation occurs under stones and at the water's edge.

Common Green and Brown Lacewings, which belong to a closely related order, the Neuroptera, are sometimes modeled by fly-fishers. The larvae of lacewings are terrestrial and eat soft-bodied insects such as aphids.

ALDERFLIES

FAMILY Sialidae
SIZE Body length up to ¾ in. Mostly under ½ in
NUMBER OF SPECIES World—100; N.A.—23; U.K.—2; Aus—4

These dull, dark insects are found resting on waterside vegetation. The adults are day-flying but spend most of their time resting. The larvae are aquatic and can be found under stones in muddy-bottomed ponds, canals and slow streams.

Female alderflies lay compact masses of eggs on plants. The hatching larvae drop or wriggle into the water (Sialidae).

The larvae, which have powerful jaws, eat small aquatic insects and worms. Females lay egg masses on vegetation or stones very close to water. Hatching larvae fall or crawl into the water. The larval abdomen has a terminal gill filament and seven pairs of feathery lateral gills. Pupation occurs in waterside soil, moss, or debris.

Alderfly larvae are predacious and have abdominal gill filaments (Sialidae).

DOBSONFLIES & FISHFLIES

FAMILY Corydalidae
SIZE Wingspan up to 6 in
NUMBER OF SPECIES World—100; N.A.—20; U.K.—0; Aus—22

This family has no representatives in Europe. Corydalids occur in almost every sort of freshwater, even temporary streams, ponds, and swamps. Some species can be large and, in some, the males possess huge jaws.

The larvae are aquatic, preferring streams with clear water and stony or muddy bottoms. Sometimes known as hellgrammites, young dobsonflies are extremely fierce predators attacking worms, crustaceans, snails, and other insects.

THE ARTIFICIAL FLIES

ALDERFLY LARVA

DRESSING

HOOK: TMC 100, size 10

THREAD: Waxed brown gossamer

TAIL: Medium brown cock hackle

BODY: Underbody—lead wire; Overbody—stripped peacock herl

THORAX: Black ostrich

GILLS AND LEGS: Short grouse hackle laid over back and tied stem at head and tail of body before thorax is made

---

FISHING METHOD

Alder larvae live on the bottom where they prey on smaller animals. Fish along the bottom, moving the fly slowly and smoothly at about ¾ in per second. Every now and then dart the fly forward (and, if using a floating line, upward) a few inches to imitate the attack of prey or an attempt to escape from a predator. Such movements may induce a following trout to take.

ANGLERS' NAME
SWANNUNDAZE ALDER LARVA

DRESSING

HOOK: TMC 5262, size 10

THREAD: Brown

TAIL: Red game hackles

BODY: Underbody—silver flat tinsel;

Overbody—Swannundaze

THORAX: Orange seal fur

WINGCASE: Pheasant tail

HACKLE: Gray partridge

FISHING METHOD

A useful imitation of the voracious alderfly larva. The translucent abdomen is particularly effective, at least to our eyes. Fish slowly along the bottom, darting it along over short distances on occasions.

ANGLERS' NAME
BLACK PALMER

DRESSING

HOOK: TMC 100, size 12	**RIB:** Silver wire
THREAD: Black	**HACKLE:** Black hen or cock
BODY: Black dubbing blend	palmered

FISHING METHOD

This pattern can be used to imitate a dead or emerging alderfly when fished in the surface film or near the surface, or if wound with a stiff cock hackle, an adult sitting on the water. Either leave it to float with the current or impart appropriate movement (as for a stonefly or mayfly adult).

ANGLERS' NAME
ADULT ALDERFLY

DRESSING

HOOK: TMC 100, size 12	**WINGS:** Paired grouse tied back
THREAD: Black	wet-style
BODY: Black goose	**HACKLE:** Black hen

FISHING METHOD

This is a wet pattern which may be used for a struggling or drowned adult alderfly.

BEETLES

THE NATURAL INSECTS

ORDER Coleoptera
DERIVATION Greek: *koleos*–sheath; *pteron*–a wing
SIZE Body length up to 7 in. Mostly under 1 in
NUMBER OF SPECIES World —370,000; N.A. —24,000;
U.K. —3,730; Aus —28,000

Beetles are generally tough-bodied and can be found everywhere from the equator to the polar regions and in every conceivable terrestrial and freshwater habitat.

The head bears forward-pointing, biting mouthparts, usually a pair of well-developed compound eyes and antennae of very variable appearance. The most distinctive beetle characteristic is the front pair of wings, or elytra, which are toughened and act as wing cases to protect the much larger, membranous hind wings folded underneath. The legs are modified to serve a variety of functions such as running, digging, jumping, and, in aquatic species, swimming. The majority of species are herbivorous but there are many scavengers, predators and a few parasites.

A ladybird taking off shows the hind wings projecting from beneath the hardened front wing cases or elytra.

If abundant, ladybirds are readily eaten by trout.

LARVAE

Beetle larvae, like adults, differ widely between species, but all have toughened heads with biting mouthparts and short antennae. They typically have three pairs of thoracic legs although these may be reduced or absent in some species.

Some 30 or so beetle families contain aquatic or partially aquatic species but only one is considered here as worthy of mention, at least as far as trout food is concerned. In some freak years, certain beetles such as ladybugs can become so abundant locally or regionally that fish eat large numbers.

Whirligig beetles swim about rapidly on the surface and look as if they have no legs. These beetles are distasteful to trout (Gyrinidae).

DIVING BEETLES

FAMILY Dytiscidae
SIZE Body length 1½ in. Mostly up to 1 in
NUMBER OF SPECIES World—3,250; N.A—475; U.K—110; Aus—200

Diving beetles, or predacious diving beetles as they are also known, are oval in outline and very streamlined with both the top and under sides convex, smooth, and shiny. The general body color is black or dark brown but many species have yellow, green, or brown bands, spots, and other patterns. The head, which seems partly sunk into the thorax, has a pair of compound eyes and threadlike antennae. The paddle-like hind legs, which are flattened with fringes of hairs, are kicked together to provide propulsion.

A large male diving beetle in search of prey (Dytiscidae).

Diving beetles are found in streams, ditches, canals, lakes and ponds, usually preferring shallower water. Some species are found in brackish water and a couple of species have even been found living in thermal springs. Adults and larvae are highly predacious on a wide range of aquatic organisms ranging from other insects to snails, tadpoles, frogs, newts, and small fish. The adults can remain submerged for long periods as they carry a supply of air trapped below their wing cases. The air store is periodically renewed by the beetles coming to

The larvae of diving beetles, which are highly predacious, are often called water tigers (Dytiscidae).

the surface rear-end first. Although these beetles are highly modified for aquatic life they fly well and do so generally at dusk and during the hours of darkness. Adults have defensive glands behind the head and at the end of the abdomen, whose secretions protect them from many predacious fish. Eggs are usually laid singly inside plant stems.

Due to their highly predacious nature, the larvae are often called water tigers and are elongate with three pairs of well-developed, hairy legs and large, curved jaws. Larvae obtain air from the surface, although there may be pairs of gill present on the sides of the abdomen. Water tigers will attack prey much bigger than themselves, sucking the body contents of their victims through grooved mandibles, after first pumping in a mixture of digestive enzymes. After three larval stages, pupation takes place in wet soil, under stones, logs, and debris close to water.

Dytiscids can be pests in garden ponds, fish hatcheries, and the like but are generally of benefit elsewhere in that they eat huge numbers of larval midges and mosquitoes.

THE ARTIFICIAL FLIES

QUICK SIGHT BEETLE

DRESSING

HOOK: TMC 100; sizes 12–16	with a spot of fluorescent orange paint to aid visibility
THREAD: Black	
BODY: Black closed cell foam	LEGS: Thick black thread
	HEAD: Black closed cell foam

FISHING METHOD

Terrestrial beetles are not noted for their prowess in the water. Present the fly on the surface, allowing it to splash down, and impart small movements to represent the drowning beetle.

COCKCHAFER

DRESSING

HOOK: TMC 100, size 12

THREAD: Black

TAIL: Five heron or goose herl fibers (natural gray)

BODY: Palmered light red game hackle

RIB: Fine gold wire

WINGCASE: Pheasant breast feather tied delta-style (flat over back)

FISHING METHOD

Cockchafers and their relatives are rather large beetles, which

are not at home in water. Having crash-landed they proceed to struggle and drown without any real success in swimming. Cast the fly onto the surface with a splash and twitch it.

FLASHBACK BEETLE

DRESSING

HOOK: TMC 3769, size 12–16

THREAD: Black

BODY: Brown floss

WINGCASE: Blue shellback material

LEGS: Black cock hackle

FISHING METHOD

As for other small beetles, fish on the surface and imitate the helpless twitching of the drowning insect.

ANGLERS' NAME
DYTISCUS LARVA

DRESSING

HOOK: TMC 300, size 6–8

THREAD: Brown

TAIL: Red game

BODY: Natural latex

THORAX: Ginger dubbing blend

WINGCASE: Cinnamon turkey herl

LEGS: Brown partridge

FISHING METHOD

A useful imitation of a dytiscid larva. Fish the fly slowly along the bottom with occasional darting movements.

ANGLERS' NAME
DYTISCUS BEETLE

DRESSING

HOOK: TMC 100, size 6

THREAD: Black

BODY: Cream yellow wool

RIB: Black silk or oval gold tinsel

THORAX: Cream yellow wool

WINGCASE: Olive goose or swan wing feather tip (varnished)

LEGS: Olive-dyed goose wing fibers

FISHING METHOD

Like their larvae, adult dytiscids are predators that inhabit weed beds and bottom debris. They swim quickly for their size, using modified legs. Fish near the bottom and move the fly along at up to 4 in per second. Think of the beetle as it darts along hunting for prey.

TRUE FLIES

THE NATURAL INSECTS

ORDER Diptera
DERIVATION Greek: *∂i*–two; *pteron*–a wing
SIZE Body length up to 2½ in. Wingspan up to 3 in
NUMBER OF SPECIES World—119,500; N.A.—16,914;
U.K.—5,950; Aus—7,800

The true flies have a tremendous impact on mankind and are found in all aquatic and terrestrial habitats. The species comprising this very large order are easily recognized by the possession of a single pair of membranous wings. Flies are masters of flight and are able to hover, fly backward, perform 360° turns, and fly and land upside down.

Typically, flies have a mobile head with a pair of large compound eyes. The antennae may be long and threadlike or short and indistinct. The mouthparts may be adapted for piercing or stabbing but, in general, they are designed for the lapping of a wide range of liquid foods, ranging from blood and nectar to the products of plant or animal decay.

The larvae of most blow flies
feed on carrion and ∂ung.

*Like all true flies this blow fly
has a single pair of wings.*

LARVAE

Larvae, which are generally cylindrical, often elongate, grub- or worm-like, are called maggots and lack thoracic legs. Depending on species, dipteran larvae can eat just about every sort of organic material. Many are herbivores, carnivores, or parasites, while others eat decaying material, animal excrement, or fungi. Numerous species are notorious vectors of human and animal disease while others can cause losses of crops and other important plants.

Although more than 30 families contain aquatic or semi-aquatic species only six families are considered here as important in the context of fly fishing.

NONBITING MIDGES

FAMILY Chironomidae

SIZE Body length ¼ in

NUMBER OF SPECIES World—4,000; N.A.—820; U.K.—400; Aus—200

Chironomids are pale green, brown, or gray gnatlike flies. They resemble mosquitoes but have no wing scales and have very small mouthparts. The males have very feathery antennae and a slender body whereas females have hairy antennae and a stoutish body. The legs are long and bear numerous fairly long hairs and bristles. The elongate wings lie flat or are held slightly raised over the abdomen.

A recently emerged midge waits for its wings to harden before it flies off

Chironomids are very common everywhere and can frequently be seen in large swarms at dusk around ponds, lakes, and streams. Larval stages can be found in virtually all types of wet habitat from bogs, wet soil, and moss clumps to lakes, ponds, streams, and rivers. The adults do not bite and survive only for a week or two after emergence. Male chironomids form mass mating swarms and

A mating swarm of nonbiting midges (Chironomidae).

122

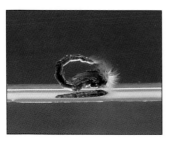

A Chironomid pupa
(Chironomidae).

The larvae of some chironomid midges are called bloodworms due to the presence of red respiratory pigment **(Chironomidae).**

mate with the females on the wing, a process that typically takes only a few seconds. The majority of the two or three year life cycle is spent as a larva. Mating occurs in swarms and the eggs are dropped on the water's surface or laid in a jelly-like substance on vegetation. The slender, green, brown, whitish, yellowish, or red larvae, which may be anything from up to 1¼ in long, feed on fine particles of decayed organic material, algae, or plant material. Some species are predacious or parasitic on aquatic invertebrates. In suitable conditions there may be thousands of chironomid larvae in a single square yard. The larvae of some species that live in oxygen-poor or stagnant water are called blood worms. Their red coloration is due to the presence of hemoglobin in the body fluids. Chironomid larvae either live on the bottom among debris inside small tubes made of a gelatinous substance coated with silt particles or swim freely. When fully grown the bottom-living species pupate inside their tubes whereas other species continue to swim free using anal lobes and hair fringes. Pupae are called buzzers by some, whereas others use the term for all chironomid life stages. In temperate regions there can be four generations in a year but one or two cycles is more common.

A few species can be pests but the majority are an important source of food for many aquatic animals. Of all the fly groups, these are by far the most important as a source of food for freshwater fish. Chironomid flies are often used as indicators of water quality as species vary in their tolerance of pollution.

MOSQUITOES

FAMILY Culicidae
SIZE Body length up to ½ in
NUMBER OF SPECIES World—3,100; N.A.—150; U.K.—36; Aus—280

These very slender, delicate flies have small, roundish heads with large compound eyes and slender, sucking mouthparts. The antennae are feathery in males and slightly hairy in females. The wings are long and narrow with scales along the veins and margins and the legs are very slender.

Mosquitoes, although found everywhere from polar regions to deserts, are most abundant in the world's warmer regions. Adults, which are common at dusk in shaded woodland and forest, do not tend to travel far from the larval breeding grounds. Larval mosquitoes occur in virtually every aquatic habitat. Female mosquitoes are blood suckers and will attack a wide range of vertebrate host animals. Males do not bite but occasionally feed on nectar or honeydew. Following mating on the wing, eggs are laid singly or in floating rafts of 40–300 eggs on the surface of water. Any water will do, from rain-filled containers and tree holes to ponds and lakes.

Mosquito larvae can often be seen hanging just below the water film (Culicidae).

*Recognisable by its feathery antennae,
a male mosquito rests on the water
surface* (Culicidae).

The larvae, which are called wrigglers from the way they thrash about in the water, mostly feed on detritus and decaying organic material, although some are predacious. In most species, air is obtained from the surface through breathing holes or spiracles at the end of the abdomen. Some larvae get their oxygen from aquatic plants by means of special sharp siphon device. The life cycle, which involves four larval stages and a pupal stage, takes 15–20 days.

Many mosquitoes are very serious disease vectors in many parts of the world. Among the human diseases they transmit are malaria, yellow fever, dengue, filariasis, and encephalitis. The same artificial flies used for chironomids can mimic mosquitoes.

*The eyes, wings and antennae of the
adult can be seen through the skin of this
mosquito pupa* (Culicidae).

THE ARTIFICIAL FLIES

MARABOU BLOODWORM

DRESSING

HOOK: TMC 100, Size 14	**BODY:** Red floss
THREAD: Black	**RIB:** Fluorescent red silk
TAIL: Red marabou	**HEAD:** Peacock herl

FISHING METHOD

The larvae of some chironomid midges contain hemoglobin, which gives them a distinctive blood-red color. They live on the bottom where they feed on detritus. Fish the fly deep and retrieve it very slowly (under ½ in per second), all the while quivering your hand. This will cause the marabout tail to wiggle and undulate like the live larva. As for all marabou-tailed flies, the tail looks fluffy and distinctly unnatural when dry but becomes sleek and alive when wet.

ANGLERS' NAME
PHANTOM LARVA

DRESSING

HOOK: TMC 100, size 12–16	Overbody—stretched clear
THREAD: Black	polythene
BODY: Underbody—silver tinsel;	**THORAX:** Orange floss

FISHING METHOD
Fish extremely slowly along
the bottom.

ANGLERS' NAME
PHANTOM PUPA

DRESSING

HOOK: TMC 100, size 14	**RIB:** Flat fine silver
THREAD: Black	**THORAX:** Orange ostrich
BODY: Yellow floss	**WINGCASE:** Slip of pheasant tail

FISHING METHOD
This pattern is a good pupal
imitation for midges and
mosquitoes. Fish it in the
surface film and either leave
it to float there or jiggle it
almost imperceptibly to represent the wriggling insect.

BRASSIE

DRESSING

HOOK: TMC 100; sizes 14–20   **BODY:** Copper wire

THREAD: Black   **THORAX:** Peacock herl

FISHING METHOD

Another useful imitation of a midge or mosquito pupa about to hatch (or actually just emerging). Fish in the surface film, leaving it stationary or twitching it very slightly.

ANGLERS' NAME
HATCHING MIDGE PUPA

DRESSING

HOOK: TMC 100, size 12–16   **THORAX:** Red floss

THREAD: Black   **HEAD:** Peacock herl

TAIL: White Antron fibers   **BREATHING SIPHONS:** White

BODY: Black floss tied sparse   antron floss

RIB: Silver wire

FISHING METHOD

A beautiful mimic of a midge pupa about to hatch. Fish in the surface film, perhaps retrieving extremely slowly. Hold the

line and try (if indeed you need to try) to get your hand to quiver as it might after several cups of strong coffee. This will impart a wriggling movement to the fly.

ANGLERS' NAME
CARNHILL'S ADULT BUZZER
(EMERGER)

DRESSING

HOOK: TMC 100, size 12–18 | **THORAX:** Mole
THREAD: Black | **EMERGING WINGS:** Two white
BODY: Olive dubbing blend | hackle tips
RIB: Silver wire

FISHING METHOD
Another emerging midge or mosquito which should be presented in the surface film.

ANGLERS' NAME
SUSPENDER MIDGE

DRESSING

HOOK: TMC 100, size 14 | **RIB:** Gold wire
THREAD: Black | **HEAD:** Bronze peacock herl
TAIL: White Antron fibers | **SUSPENDER BALL:** White
BODY: Black dubbing blend | plastazote in nylon tights

FISHING METHOD
This bears a strong resemblance to a midge bursting free of its pupal exuvium. Fish suspended in the surface film.

OLIVE MIDGE

(EMERGER)

DRESSING

HOOK: TMC 100, size 12

THREAD: Pale primrose

BODY: Golden olive dubbing blend

RIB: Yellow floss

WINGS: Two blue dun hackle points tied back over body

HACKLE: Two turns of grizzly

FISHING METHOD

This tangled-looking fly can be used to imitate an emerging midge or caddisfly, if fished in the surface film and agitated slightly.

OLIVE MIDGE

DRESSING

HOOK: TMC 3769, size 18

THREAD: Olive

BODY: Olive fur

HACKLE: Grizzly cock

FISHING METHOD

A useful dry fly to represent any very small winged insect, such as a midge. Fish on the surface without imparting any movement.

ANGLERS' NAME
GRAY DUSTER

DRESSING

HOOK: TMC 3769, size 12	**BODY:** Gray antron
THREAD: Black	**HACKLE:** Badger cock

FISHING METHOD
A bushy little dry fly which is a useful imitation of many small winged insects, including mosquitoes and midges. Treat with floatant and leave it to drift on the surface.

ANGLERS' NAME
REED SMUT

DRESSING

HOOK: TMC 3769, size 18	**BODY:** Black seal fur
THREAD: Black	**HACKLE:** Grizzly cock

FISHING METHOD
A very effective imitation of an adult midge, or indeed any tiny winged insect. Treat with floatant and fish on the surface. There is no need to impart any but the slightest movement.

DOUBLE BADGER

DRESSING

HOOK: TMC 100, size 14
THREAD: Black
BODY: Bronze peacock herl

WINGS: Two wound badger cock hackles

FISHING METHOD

Adult midges have only one biological aim: to reproduce. A mating pair on the water surface is a bonus to a trout: two for the price of one. Cast the fly onto the surface and leave it there.

MOSQUITO ADULT

DRESSING

HOOK: TMC 100, size 14–20
THREAD: Black
BODY: Thinly dubbed gray polypropylene

RIB: Black silk
WINGS: Two small grizzly hackles tied spent
HACKLE: Blue dun cock

FISHING METHOD

This fly can be used to represent any gray-colored midge or mosquito, or even a mayfly. Fish on the surface and move only slightly, if at all.

ANGLERS' NAME
GRIFFITH'S GNAT

DRESSING

HOOK: TMC 100, sizes 16–20 | wound the entire length of the
THREAD: Black | body
BODY: Peacock herl
HACKLE: Grizzle cock hackle

FISHING METHOD

A simple, yet effective, imitation of a small, black midge. Present on the water surface and impart no movement.

CRANE FLIES

FAMILY Tipulidae
SIZE Body length up to 2½ in. Mostly up to 1 in
NUMBER OF SPECIES World—14,000; N.A.—1,525; U.K.—318; Aus—700

A crane fly in flight (Tipulidae).

Crane flies (also known as daddy-longlegs) are brown, black, or gray, often with yellow, orange, or pale brown markings. The body is elongate and slender with a distinctive "V" shaped groove on top of the thorax. The wings are long and often have dark markings or spots. The head, which carries relatively long antennae, is prolonged in front of the eyes to form a short rostrum or beaklike structure bearing the mouthparts. A characteristic feature of flies belonging to this family is that their legs, which are very long and thin, are shed very easily if they are trapped or handled. The end of the abdomen is blunt and expanded in males whereas females have a sharply pointed, tough ovipositor.

Adults are commonly found near water or among grass-land and rank vegetation. Many woodland species are common and very abundant. Eggs are laid in the soil or in debris and the elongate, cylindrical larvae, which are mostly brown or gray in color, may be aquatic, partially aquatic, or terrestrial in habits. At certain times of the year, crane flies may become very abundant and commonly land on or are blown onto water. Some species live for only a few days as adults and may not feed at all while other species take nectar or similar fluid foods. The elongate, cylindrical larvae, which are not nearly as well known as the adults, are known as leather-jackets on account of the tough texture of their bodies. The larvae, which all require a certain degree of moisture, live in soil, rotting wood, fresh water, swamps, and bird nests where they eat plant and animal material or detritus. The larvae of the truly aquatic species live submerged all the time and take in oxygen through their cuticle. The adults of many species of crane fly are readily attracted to sources of light at night.

Giant crane fly (Tipula maxima) hanging from a blade of grass.

After mating, this female crane fly (on the left) will probably lay her eggs in moist soil (Tipulidae).

At certain times of the year crane flies can be abundant in vegetation near water (Tipulidae).

The larvae of several species are pests of grasses, crops, and garden plants, damaging the plants' roots. Crane flies and their larvae, being the food of many vertebrate and invertebrate animals, are an important component of terrestrial and aquatic food chains.

The larvae of crane flies live in damp soil or water (Tipulidae).

THE ARTIFICIAL FLIES

ANGLERS' NAME
CRANE FLY LARVA

DRESSING

HOOK: TMC 5262, size 8	latex
THREAD: Brown	**RIB:** Oval gold tinsel
BODY: Underbody—wool or silk;	**HACKLE:** Small clipped white
Overbody—yellow dyed dental	hackle

FISHING METHOD

The larvae (leatherjackets) of some crane flies are aquatic or semiaquatic, living in the mud at the bottom and sides of the water. Having no legs they do not move far or fast (although some species can swim by flattening the body). Fish very slowly along the bottom, close to the water margins.

ANGLERS' NAME
DADDY-LONGLEGS

DRESSING

HOOK: TMC 5262, size 10	**LEGS:** Six pheasant tail fibers
THREAD: Brown	knotted twice tied trailing
BODY: Pheasant tail fibers	**HACKLE:** One large Furnace
WINGS: Two pairs red game	hackle
hackles tied spent	

FISHING METHOD

Yet another excellent imitation of a drowned adult crane fly, to be fished near the surface.

DADDY-LONGLEGS

(WALKER PATTERN)

DRESSING

HOOK: TMC 5262, size 10

THREAD: Brown

BODY: Varnished natural raffia

WINGS: Four chinchilla grizzle hackles tied spent

LEGS: Eight pheasant tail fibers each with two knots

HACKLE: Two red game tied large

FISHING METHOD

This is a wonderful imitation of a dead crane fly. These insects

emerge in large numbers and get blown onto the water or land there by accident, where they quickly drown. Fish either on the surface or let it sink below as if waterlogged. There is no need to move the fly.

ANGLERS' NAME

ORVIS CRANE FLY

DRESSING

HOOK: TMC 100, size 14

THREAD: Black

BODY: Yellow or orange floss

HACKLE: Long black cock

FISHING METHOD

A simple but effective imitation of an orange or yellow colored crane fly adult. Fish on the surface or in the surface film to represent either a dead or drowning insect.

ANGLERS' NAME
Wet Daddy

DRESSING

HOOK: TMC 5262, size 8

THREAD: Black

BODY: Natural raffia

RIB: Fine gold wire

SEMI-HACKLE: Light red-brown cock palmered halfway

down body to look like legs (optional)

HACKLE: Long brown partridge fibers over golden pheasant tippet

FISHING METHOD

One of the simplest dead crane fly imitations, which is effective if allowed to sink below the surface and drift with the current. A very slow retrieve can be used to augment natural drift and cause the soft hackle to open and close enticingly.

ANGLERS' NAME
Detached Body Crane Fly

DRESSING

HOOK: TMC 100, size 10

THREAD: Brown

TAIL: Elk or deer hair detached and bound with white thread

WINGS: Two chinchilla grizzly tied spent

LEGS: Six pheasant tail fibers knotted twice

HACKLE: Two red game cock tied long

FISHING METHOD

This pattern is a different style of crane fly adult. Fish on the surface or in the surface film, either as a dead insect or one that is struggling vainly to escape drowning.

MARCH OR HAWTHORN FLIES

FAMILY Bibionidae
SIZE Body length up to ½ in
NUMBER OF SPECIES World—780; N.A.—78; U.K.—19; Aus—32

March flies are stout-bodied, black or dark brown insects, often with hairy bodies and shortish legs. The heads of males are large and have well-developed compound eyes that meet on top of the head. Females have smaller heads and eyes that

March flies are most common in spring and early summer **(Bibionidae).**

do not meet. Both sexes have short, beadlike antennae. The wings are large and have a pale brown tinge.

March flies are common on flowers in pastures, gardens, and similar habitats during spring and early summer. Large mating swarms of these flies can be seen in spring and often end up on the surface of water where they are taken avidly by trout. Mated females dig down through the soil and lay 200–300 eggs in a small chamber. The scavenging larvae, which are elongate, slightly flattened with a large head and strong mouthparts, eat all manner of organic material including plant roots. Larvae can be found in rich soil, compost heaps, dung, leaf litter, and meadows. Pupation takes place underground in an earthen cell.

A pair of march flies mate on the surface of a leaf **(Bibionidae).**

THE ARTIFICIAL FLIES

HAWTHORN FLY

DRESSING

HOOK: TMC 100, size 12–14

THREAD: Black

BODY: Black polypropylene

WINGS: Light blue dun hackle points

LEGS: Two strands of black horsehair trailing

HACKLE: Black cock

FISHING METHOD

Adult bibionids are abundant for short periods and often fall into the water where they drown. This pattern should be fished on the surface or allowed to sink. Move only sufficiently to indicate the insect's dying struggles.

HOVER OR FLOWER FLIES

FAMILY Syrphidae
SIZE Body length up to 1¼ in. Mostly up to ¾ in
NUMBER OF SPECIES World—6,000; N.A.—870; U.K.—245; Aus—170

Hover flies are among the most easily recognizable and common species of all the Diptera. The adults can be blue, black, or metallic but many have a bee or wasplike appearance with yellow stripes, spots, or bands.

Although hover fly species can be very variable in size and coloring they are most easily spotted in the field by the way they hover and dart between flowers. Syrphids are found at a large variety of flowers in a wide range of terrestrial habitats. Adult hover flies feed on pollen and nectar while their larvae may eat plants or soft-bodied insects such as aphids and scale insects. Detritus-eating species live in rotting wood, dung, mud, or dirty water. The larvae of some of the aquatic species such as those of the Drone Fly have an extensible posterior siphon through which they can get air from the surface, giving them the name rat-tailed maggot.

Many hover flies such as this Drone Fly are good mimics of bees or wasps (Syrphidae).

THE ARTIFICIAL FLIES

ANGLERS' NAME
RAT-TAILED MAGGOT

DRESSING

HOOK: TMC 100, size 12
THREAD: Brown
TAIL: Undyed swan feather fibers

BODY: Fluorescent white wool
RIB: Buff cock hackle stalk
HEAD: Two turns of medium brown ostrich herl

FISHING METHOD

Rat-tailed maggots are the larvae of the drone fly (*Eristalis*). They live on the bottom in shallow water near the margins of pools. They have a long breathing tube at the rear end which is imitated in this pattern. Fish close to the bank along the bottom and move only very slowly (under ½ in per second).

GRAFHAM DRONE

DRESSING

HOOK: TMC 100, size 14

THREAD: Black

BODY: Yellow floss

RIB: Black thread

WINGS: Bunch of blue dun hackle fibers

HACKLE: Red game cock hackle tied short

FISHING METHOD

Drone flies may have aquatic larvae (the rat-tailed maggot) but they cannot swim. Fish the fly on the surface making it struggle as if drowning, or let it sink and leave it to drift along.

ANGLERS' NAME
DRONE FLY

DRESSING

HOOK: TMC 100, size 14

THREAD: Red

BODY: Wound yellow ostrich herl

RIB: Black ostrich herl

WINGS: Two white hackle tips

HACKLE: Natural red game cock

FISHING METHOD

Another imitation of the drone fly, which should be fished on the surface or in the surface film.

ANGLERS' NAME
BLUEBOTTLE

DRESSING

HOOK: TMC 100, size 12–14
THREAD: Black
BODY: Blue flashback wrapped with black ostrich herl

WINGS: Two blue dun hackle points tied flat
HACKLE: Black cock

FISHING METHOD
Bluebottles cannot swim, so fish this pattern on or just under the surface, letting it just float along or jiggle it slightly.

ANGLERS' NAME
BLACK GNAT

DRESSING

HOOK: TMC 100, sizes 12–20
THREAD: Black
TAIL: Black cock hackle fibers
BODY: Black wool

WING: Paired slips of gray mallard primary
HACKLE: Dyed black cock

FISHING METHOD
Fish as for the bluebottle.

CADDISFLIES

THE NATURAL INSECTS

ORDER Trichoptera
DERIVATION Greek: *trichos*–hair; *pteron*–a wing
SIZE Body length up to 1½ in
NUMBER OF SPECIES World—9,000; N.A.—1,265; U.K.—192; Aus—480

Caddisflies can be found anywhere there is fresh water. The drab-colored adults are mothlike but unlike lepidopterans the body and wings are covered with hairs not scales. Another distinguishing feature is that caddisflies do not have a coiled proboscis. The head has a pair of compound eyes and long, threadlike antennae. Although the weakly developed mouthparts can be used for drinking water and nectar, the adults of many species do not feed. The head and parts of the thorax may have small, warty bumps. The wings are held over the body in a tentlike fashion and the legs are long and quite slender.

Caddisflies can be moth-like but the wings are not scaly and are held roof-like over the body.

Mostly nocturnal in habits, the adults hide in vegetation during the hours of daylight and are quite hard to find. Mating takes place at dusk, either in flight (sometimes in swarms) or on vegetation. The egg masses or strings, which are covered in a gelatinous substance, are dropped onto the water, laid under water or left on overhanging objects from which they will eventually fall.

Caddisflies mating on foliage overhanging water.

LARVAE

Caddisflies are better known for the beautifully constructed cases inside which the aquatic larvae live. Caddis larvae are caterpillar-like with a strong head and thorax and a soft abdomen. The head has chewing mouthparts and short antennae. The thorax bears three pairs of legs and the end of the abdomen has a pair of hooked prolegs to anchor the larva in its case. Oxygen is obtained by means of tufts of external gills located on the abdominal segments. Caddis larvae may be grazers, predators, or scavengers. Not all larvae make cases, some are free-living while others make silken nets to filter minute organic particles of food from the water currents. The cases, which are open at both ends, are added to at the front as the occupant grows. They are built from a

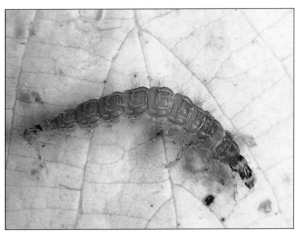

The larvae of some caddisflies do not make a portable case but live free on the bottom.

The larval case of this caddisfly has been made out of leaf fragments.

great variety of materials such as plant twigs, stems and leaves, snail shells, sand grains, shingle, and small stones, which are held together with silk secreted from glands in the head. The case design and construction is specific to particular caddis families and a useful identification feature. In species that exclusively use twigs as construction materials, the cases are called log cabins. The larvae of some species use larger twigs which stick out at angles and make it difficult for fish to swallow them.

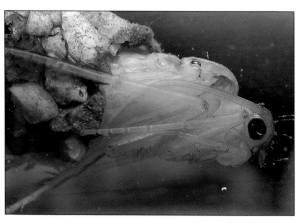

*A caddisfly pupa emerging from a larval case
made from small pieces of gravel.*

After five larval stages, pupation takes place within the silk-sealed larval case or in a specially made pupal case which is firmly attached to the substrate. The pupa has sharp mandibles which it uses to cut its way out of the case. Adults, still enveloped in the pupal skin, swim upwards and molt at the surface. The emerging adults need to expand and harden their wings before taking to the air. In temperate regions there is only one generation per year. All caddisfly life history stages are eaten by many freshwater fish and birds.

An adult caddisfly resting on a twig.

THE ARTIFICIAL FLIES

CADDIS LARVA

DRESSING

HOOK: TMC 5262, size 10	**RIB:** Fine gold wire
THREAD: Black	**THORAX:** White floss or silk
BODY: Dubbed hare's ear	**HACKLE:** Short black hen

FISHING METHOD

Caddis larvae live on the bottom and move only slowly. Fish the fly deep and retrieve smoothly at about ½ in per second, pausing occasionally.

CASED CADDIS

DRESSING

HOOK: TMC 800 or TMC 5262, size 6–12	medium grains of gravel stuck to wet epoxy adhesive and dried
THREAD: Brown	**THORAX:** White ostrich herl
BODY: Underbody of wool with	**HACKLE:** Black hen

FISHING METHOD

This is a particularly lifelike imitation of those caddisfly

larvae, which incorporate gravel in their "case." Other items such as small pieces of stick can also be used to represent other species. Fish the pattern very slowly and smoothly along the bottom.

ANGLERS' NAME
STICKFLY

DRESSING

HOOK: TMC 5262, size 10
THREAD: Brown
BODY: Pheasant tail
RIB: Copper wire
THORAX: Fluorescent yellow floss
HACKLE: Short red game cock

FISHING METHOD
Another well-tried imitation of a caddis larva, to be fished very slowly along the bottom.

ANGLERS' NAME
GREEN TAG STICKFLY

DRESSING

HOOK: TMC 100, size 12
THREAD: Brown
TAG: Fluorescent green floss
BODY: Bronze peacock herl
HACKLE: Dark red game

FISHING METHOD
Fish slowly and deep as for other caddis larva imitations.

LONGHORNS

DRESSING

HOOK: TMC 100, size 12

THREAD: Black

BODY: Amber dubbing on first two thirds with dark brown on last third

WINGCASE: Pheasant tail fibers tied back

ANTENNAE: Two pheasant tail fibers

HACKLE: Short red game

FISHING METHOD

A marvelous imitation of a caddis pupa, with its long antennae. Caddis pupae rise rapidly to the surface when ready to emerge. Fish the pattern in the surface film, and agitate to indicate the struggle to escape quickly from the pupal exuvium (shuck). If the pattern is weighted and fished on a floating line it can be left to sink then caused to rise up to the surface by pulling smoothly on the line. Let the fly sink again and repeat the retrieve. Be ready for savage attacks.

ANGLERS' NAME
GREEN DF PARTRIDGE

DRESSING

HOOK: TMC 100, size 12

THREAD: Black

BODY: Fluorescent green floss

RIB: Fine oval silver wire

HACKLE AND LEGS: Partridge hackle

FISHING METHOD

Another pattern which can be used as a mimic of a caddis pupa. Vary the color of tying and keep a selection of brown,

green, black, yellow, and green versions. Fish either in the surface film or, using a floating line, allow to sink and then retrieve, as if it is heading to the surface to emerge.

ANGLERS' NAME
BEAD HEAD EMERGING CADDIS

DRESSING

HOOK: TMC 3761, sizes 10–14

THREAD: Gray

TAIL: A few strands of Sparkle yarn

UNDERBODY: Gray fur and brown sparkle yarn mixed

OVERBODY: Gray Sparkle yarn pulled loosely over the body

WING: A few strands of gray deer hair

THORAX: Peacock herl

HEAD: Gold bead

FISHING METHOD

This pattern can be used to imitate certain caddis pupae as well as midges and mosquitoes. Fish in the surface film or rising up from the bottom as for other caddis pupae.

BLACK SEDGE

DRESSING

HOOK: TMC 100, size 10

THREAD: Black

BODY: Black wool or chenille

WINGS: Black deer or elk hair
tied flat and cut square

HACKLE: Black cock tied to
slope forward

FISHING METHOD

Adult caddisflies are agile fliers. After emerging from their pupal exuvium they sit on the water to inflate and dry their wings then skitter across the surface into flight. Mated females of most species oviposit directly into the water by skimming over the surface. This fly pattern can be used to represent any one or all of these behaviour patterns. Treat it with floatant, cast it onto the surface and leave it sit for a while. Then pull the fly across the water so that it accelerates, covering a yard or so in about three seconds. Let it rest again (takeoff aborted) and repeat the retrieve. Egg laying can be imitated by moving the fly at a constant rate of 4 in per second over several yards, then accelerating it as if it is about to take flight. Be ready for some explosive attacks.

ANGLERS' NAME
ELK HAIR CADDIS

DRESSING

HOOK: TMC 100, sizes 10–18

THREAD: Tan

BODY: Hare's fur either natural or dyed olive or black ribbed with fine gold wire

HACKLE: Brown cock hackle wound over body

WING: Bleached elk hair

FISHING METHOD

This pattern is an excellent imitation of an adult caddisfly. Fish as for the Black Sedge.

ANGLERS' NAME
LARGE BROWN SEDGE

DRESSING

HOOK: TMC 5262, size 10

THREAD: Orange

TAG: Yellow fluorescent floss

BODY: Clipped chestnut ostrich herl

WINGS: Bunch of red game hackles tied flat over back

HACKLE: Red game cock

FISHING METHOD
Fish as for the Black Sedge.

ANGLERS' NAME
GRANNOM

DRESSING

HOOK: 79704 BR, size 16

THREAD: Green

TAG: Fluorescent green wool

BODY: Natural heron herl

WINGS: Blue dun cock fibers

HACKLE: Ginger cock

FISHING METHOD
A very effective imitation of smaller, rusty brown caddis species. Fish dry, interspersing periods of quiet drift with scoots of 20 in across the surface.

ANGLERS' NAME
STOCKING SEDGE

DRESSING

HOOK: TMC 100, sizes 14–18

THREAD: Brown

BODY: Yellow wool with a brown cock hackle wound over

WING: Light mottled turkey tail stuck to a sheet of stocking mesh and then cut to shape

HACKLE: Brown cock hackle tied full

ANTENNAE: Two cock pheasant tail fibers

FISHING METHOD
An excellent mimic of caddis adults with long antennae.

ANGLERS' NAME
PALE SEDGE

DRESSING

HOOK: TMC 100, size 10

THREAD: Brown

BODY: Cinnamon turkey tail

RIB: Gold twist

WINGS: Natural hen pheasant wing fibers rolled and tied flat

HACKLE: Ginger cock at head and palmered ginger cock over

FISHING METHOD

Very usefully employed as a caddis adult resting on the surface prior to taking flight. Leave it drift with the current and every now and then scoot it across the surface for about 20 in.

ANGLERS' NAME
CINNAMON SEDGE

DRESSING

HOOK: TMC 100, size 10

THREAD: Black

BODY: Brown pheasant tail fiber

WINGS: Cinnamon turkey hackle

HACKLE: Ginger or natural red cock

FISHING METHOD

A standard pattern for imitating reddish-brown caddis adults sitting on the surface. Treat with floatant and let the fly drift along and occasionally pull to indicate the insect accelerating before takeoff.

ANGLERS' NAME
CAPERER

DRESSING

HOOK: TMC 100, size 14

THREAD: Orange

BODY: Mixed orange and fiery-brown dubbing blend

RIB: Fine gold wire

WINGS: Red game cock hackles tied back

HACKLE: Red game cock

FISHING METHOD

Use to imitate an adult caddis fluttering along the surface, either laying eggs or about to take flight.

ANGLERS' NAME
G & H SEDGE

DRESSING

HOOK: TMC 5262, size 10

THREAD: Brown

BODY: Bright green floss strung from bend to head

WINGS: Spun deer hair shaped to form roof-sloped wings

ANTENNAE: Stripped red game cock hackle stalks

HACKLE: Red game cock

FISHING METHOD

This buoyant caddis adult imitation has the long antennae that are such a distinctive feature of many caddisflies. Fish the fly on the surface as for other adult sedges.

158

MOTHS & BUTTERFLIES

THE NATURAL INSECTS

ORDER Lepidoptera
DERIVATION Greek: *lepidos*–scale; *pteron*–a wing
SIZE Wingspan up to 12 in. Mostly under 3 in
NUMBER OF SPECIES World — 165,000; N.A. — 11,286;
U.K. — 2,495; Aus — 21,000

Butterflies and moths form a huge order of insects second in size only to the Coleoptera. Members of the order occur everywhere there is vegetation and are generally so well known that they do not require much of an introduction.

People often think that the separation of moths and butterflies is based on scientific principles; in fact it is done more for convenience. Moths, on the whole, are nocturnal, not generally brightly colored, rest with their wings held outstretched, and do not have swollen antennal ends. Butterflies, on the other hand, are day-fliers, brightly colored, rest with their wings closed up, and have slightly swollen antennal ends. There are, of course, many exceptions and it is not impossible, for instance, to find very gaudy moths and some very drab butterflies. The most characteristic features of the order are the minute overlapping scales that cover the entire body surface and both pairs of wings. Another unique feature is the long, coiled proboscis, which is used to suck liquids. The wing scales which give these insects their fantastic colors and patterns can be either pigmented or micro-sculptured so that they refract the light falling on them.

LARVAE

Caterpillars are long and cylindrical and may be smooth, spiny, or hairy. In addition to three pairs of thoracic legs, caterpillars have a variable number of abdominal prolegs with which they hold tightly to vegetation. The prolegs have many

Most moths are nocturnal, some, like this arctiid moth, fly at dawn.

*Some caterpillars like this Garden Tiger
Moth larva are very hairy.*

tiny hooks called crochets to help them hold on to surfaces.
The vast majority of caterpillars are herbivorous and often
plant-specific but fungi, dried organic material, and lichens
are eaten by some species. Pupation takes place after
anything between four and nine larval stages. Lepidopteran
pupae may be enveloped inside a silk cocoon spun by the fully
grown caterpillar, inside an underground silk-lined cell, or
suspended in various ways from the food plant. The term
chrysalis is often used for lepidopteran pupae.

*Three pairs of thoracic legs and five pairs of
abdominal prolegs hold this caterpillar
firmly in place.*

Although mainly terrestrial there are many hundreds of moths species, belonging to half a dozen or so families, whose caterpillars are associated with aquatic plant and habitats. Many of these species are leaf-miners or stem-borers in water plants floating on or emerging from the surface of lakes and ponds. The caterpillars of some species may graze algae or feed on mosses or lichens and a few actually make small cases of plant material fragments around themselves just like the young of caddisflies. Aquatic caterpillars have small tufts of gills on segments of the thorax and abdomen and the papae may be anchored to the substrate by means of a hooklike terminal structure. Butterflies and moths generally only land on water as adults by accident.

A number of trout flies have been made to imitate butterflies or moths in general. Dusk-flying moths are particularly prone to ending up on water and are eaten by trout.

Noctuid moths are abundant and widespread.

THE ARTIFICIAL FLIES

ANGLERS' NAME
CATERPILLAR

DRESSING

HOOK: Swimming nymph hook, size 8–10 bent	golden pheasant tippets tied to underside in 3 evenly spaced bunches
THREAD: Brown	
BODY: Brown chenille with	**HEAD:** Bronze peacock herl

FISHING METHOD

This pattern looks remarkably like a woolly bear caterpillar that has just fallen out of an overhanging tree or bush and landed on the water, where it will wriggle around before drowning. Cast onto the surface with a plop and allow to drift along while jiggling it slightly.

ANGLERS' NAME
WAVE MOTH

DRESSING

HOOK: 79704 BR, size 16	**WINGS:** Two white goose wing slips tied upright and separated
THREAD: White	
BODY: Clipped white ostrich	**HACKLE:** White cock

FISHING METHOD

Fish this pattern as an imitation of a small white moth that is in the process of drowning or has already succumbed. It is best used at dusk when real moths are about.

GHOST SWIFT MOTH

DRESSING

HOOK: TMC 5262, size 8

THREAD: White

BODY: Cream ostrich herl

RIB: Stiff cream-colored cock hackle

WINGS: Swan secondary wing feathers

HACKLE: One cream and 1 pale ginger cock hackle

FISHING METHOD

This pattern is supposed to represent a ghost-swift moth, which is a rare insect that you are unlikely to encounter. It will however serve as a useful imitation of any large white moth. No moth can swim, so allow the fly to crash land on the surface, creating a splash, then lie there struggling before sinking under the surface. Do not drag it for long distances over the water surface, since a drowning, waterlogged moth could not achieve such a feat. Remember that most moths are only active in the evenings and after dark.

ANGLERS' NAME
HOOLET

DRESSING

HOOK: TMC 100, size 10

THREAD: Black

BODY: Cork strip with peacock herl wound over

WINGS: Cinnamon turkey

HACKLE: Two natural red cock hackles

FISHING METHOD

This pattern is usefully employed at dusk to represent any of the thousands of species of medium-sized brown moths that might crash-land on water and drown.

ANGLERS' NAME
BLACK MOTH

DRESSING

HOOK: TMC 100, size 10

THREAD: Black

BODY: Black polypropylene

THORAX: Orange fluorescent floss

WINGS: Black hen hackle tips

HACKLE: Black cock

FISHING METHOD

Use as for other moth patterns. The same pattern could also represent a black caddisfly, in which case more movement could validly be imparted to the fly.

WHITE ERMINE

DRESSING

HOOK: TMC 100, size 10

THREAD: Black

BODY: Over body—white rabbit fur or floss; Under body— orange wool or floss projecting out as a tail

RIB: Black silk

HACKLE: White cock and gray partridge

FISHING METHOD

A good imitation of an ermine moth (a species that is relatively common) but will serve for any smallish white moth which has landed on the water and drowned. Fish in the surface film with little or no movement.

HACKLE POINT COACHMAN

DRESSING

HOOK: 79704 BR, size 18

THREAD: Brown

TAG: Silver wire

BODY: Bronze peacock

WINGS: Two white hackle points tied upright and separated

HACKLE: Dark red game cock

FISHING METHOD

Another small moth imitation, this time a small brown one. Use as for other moth imitations.

SAWFLIES, WASPS, ANTS, & BEES

THE NATURAL INSECTS

ORDER Hymenoptera
DERIVATION Greek: *hymen*–membrane; *pteron*–a wing
SIZE Body length up to 2¾ in
NUMBER OF SPECIES World — 120,000; N.A. — 18,000;
U.K. — 6,650; Aus — 14,800

The Hymenoptera, which are very diverse in appearance, comprise the third largest insect order. Most hymenopterans have two pairs of membranous wings and, with the exception of the sawflies, a distinctive, narrow-waisted appearance. The head carries a pair of threadlike antennae of various designs and a pair of compound eyes. The mouthparts are generally adapted for chewing and biting but many species take liquid food. In the bees the mouthparts are highly modified to form

Bees are responsible for the pollination of almost all the world's flowering plants.

a tongue through which nectar is sucked. Sawflies are largely herbivorous, female sawflies having a sawlike ovipositor for laying their eggs inside plant tissue. Most other hymenopterans are parasites or predators of other insects and females have either a slender ovipositor for laying eggs in other insects or a modified ovipositor in the form of a sting. The larvae of sawflies are very caterpillar-like with a distinct head, thoracic legs and abdominal prolegs but the larvae of all other species are small, legless, and grublike. Pupation takes place within a silk or paperlike cocoon or in a specially made cell inside a nest. Hymenopterans are generally very useful as they destroy pest species and pollinate virtually all the known species of flowering plant on Earth.

Within the many thousands of small or minute parasitic species there are many that specialize on the eggs, larvae, or pupae of aquatic insects such as water bugs, diving beetles, and caddisflies. The winged females enter the water to locate an appropriate host, rowing down using their wings as oars. In some species the males and females may even mate hanging below the surface film. Despite being totally terrestrial in habit, bees and ants are the only hymenopterans that are regularly used as successful patterns for fly fishing.

Ants are essential components of most terrestrial ecosystems.

THE ARTIFICIAL FLIES

ANGLERS' NAME
RED ANT

DRESSING

HOOK: 79704 BR, size 14

THREAD: Red

BODY: Crimson silk with a twist of peacock herl at the tail

WINGS: Medium starling

HACKLE: Small dark ginger cock

FISHING METHOD

Winged ants are extremely abundant for short periods, usually during summer after rain. At such times they provoke frenzied feeding by trout. Since the ants cannot swim, allow the fly to land on the surface and twitch it as it drifts along. Don't worry if the fly sinks under the surface as it becomes waterlogged: that is what real ants do.

BLACK ANT

DRESSING

HOOK: TMC 100, size 14

THREAD: Black

BODY: Black acetate floss

WINGS: Two blue dun hackle

tips tied over back

HACKLE: Black cock behind

head built up from floss

FISHING METHOD
As for the red ant (see page 169).

BEE

DRESSING

HOOK: TMC 9300, size 10

THREAD: Black

BODY: Banded yellow black and white ostrich herl

WINGS: Two grizzly hackle points

LEGS: Four dyed black pheasant tail fibers

FISHING METHOD
Bees and bumblebees are not accomplished swimmers. When they end up on water they drown. Cast the fly onto the water so that it splashes down and jiggle it slightly as it drifts along.

CRAYFISH, SHRIMPS, WATERLICE, ETC.

THE NATURALS

ORDER Crustacea
DERIVATION Latin: *crusta*—rind, shell
SIZE Body width up to 14 in
NUMBER OF SPECIES World—32,000

The Crustacea, distinguished by the possession of two pairs of antennae, two-branched limbs, and, usually, gills, form a large and generally familiar group arranged in six classes within the Arthropoda (that is all animals with a hard shell and jointed legs—the insects form a further class within the Arthropoda). The vast majority of crustaceans are marine but there are many freshwater species.

The shell of crustaceans, although mostly toughened with calcium carbonate, is not topped with a thin wax layer to reduce water loss. The lack of a wax layer is one of the

Freshwater crayfish feed on snails, tadpoles and aquatic insects.

Amphipod crustaceans such as this freshwater shrimp or scud are flattened from side to side.

reasons why so few species have become completely terrestrial. Development in crustaceans depend on whether the species is marine or freshwater. Marine larval forms tend to look very different from the adult stage while those in freshwater habitats tend to look like miniature adults. Crustaceans have evolved to fill all manner of ecological niches as herbivores, predators, scavengers, and even internal and external parasites of other animals.

Within the Crustacea several orders are of importance in freshwater habitats and are used as models for fishing flies.

The Amphipoda contains some 4,000 world species of small, jumping, shrimplike species, commonly known as scuds, sand- or beach-hoppers, waterlice, and freshwater shrimps. The body appears slightly curved and is flattened from side to side.

Closely related to the amphipods, the Isopoda form a similarly sized and largely marine crustacean order. Isopods are flattened dorsoventrally and generally get around by crawling. Some isopods have evolved to fill terrestrial niches but even these species, known as woodlice, slaters, sowbugs, and pillbugs, tend to be confined to damp microhabitats. Aquatic species (also known as sowbugs or noglice) crawl in among underwater vegetation and along the bottom.

Waterfleas belong to the crustacean order Cladocera. In these, mainly freshwater species, the carapace takes the form of a bivalved shell that nearly covers the body. The 500 or so world species are small or minute and are a very important part of the diet of freshwater fish. One species, the well known *Daphnia*, is bred commercially and sold as food for aquarium fish.

Much larger than any of the preceding species, freshwater crayfish belong to the order Decapoda, that is the crustaceans with ten limbs (crabs, lobsters, shrimps, and prawns). These species, which resemble miniature lobsters, live in flowing water and feed at night on snails, tadpoles, and aquatic insects. Young crayfish appear in the spring and may spend the very early part of their lives clinging on to their mother with their small pincers.

A water slater or sowbug.

THE ARTIFICIAL FLIES

PETER GATHERCOLE'S
WATERLOUSE

DRESSING

HOOK: TMC 100, size 12–16

THREAD: Black or brown

BACK: Gray-brown feather fibers

TAIL: Some back fibers sticking out

BODY: Gray rabbit fur with a short-fibered partridge hackle over back

RIB: Silver wire

ANTENNAE: Two brown feather fibers

FISHING METHOD

Waterlice (hoglice) crawl around on the bottom and among submerged vegetation. Fish along the bottom moving smoothly and slowly at under ½ in per second.

SHRIMP

(SCUD)

DRESSING

HOOK: TMC 2487, or Yorkshire sedgehook, size 10–16	Shellback material
THREAD: Olive	**RIB:** Gold wire or oval
BACK: Clear polyethylene or	**BODY HACKLE:** Olive or orange hen or cock

FISHING METHOD

Fish along the bottom and retrieve in short pulls of ¾ in to indicate the escape hops of these creatures.

RED SPOT SHRIMP

DRESSING

HOOK: Yorkshire sedge hook, size 10

THREAD: Waxed olive

BODY: Underbody—fine lead wire; Overbody—red wool with an equal amount of olive Mohair and dubbing tied at right angles to the center of the shank/wool clipped to make spots at sides of body

BACK: Two layers of clear plastic sheet

RIB: Gold wire

LEGS: Olive body fibers picked out

FISHING METHOD

There is an interesting piece of biology behind this pattern. Water shrimps (scuds) such as *Gammarus* are parasitized by worms which cause them to develop red spots on their body and also to change their behavior. Instead of crawling around and remaining out of sight, the parasitized animals repeatedly jump up from the bottom, so attracting the attention of fish and ensuring that they are eaten. This is to the advantage of the parasite which uses the fish as the next host in its life cycle. Try to imitate this behavior by using a floating line, letting the fly sink to the bottom, and retrieving it in short jerks of about 1 in between each of which you allow the fly to sink back to the bottom. Every time you pull, the shrimp will rise up in the water a few inches.

ANGLERS' NAME
CRAYFISH

DRESSING

HOOK: TMC 300, size 4–8

THREAD: Olive

TAIL AND BACK: Yellow shellback over olive Twinkle

BODY: Palmered dyed saddle hackle

RIB: Flat Mylar medium

ANTENNAE AND MOUTHPARTS: Dyed elk hair

EYES: Black plastic beads

CLAWS: Two pairs of short dyed grizzle saddle hackles from the densest part of the hackle

FISHING METHOD

This is a beautiful imitation of a freshwater crayfish. Fish along the bottom, moving the fly along smoothly at a rate of ¾ in per second to indicate the animal walking along, then every now and then shoot it backward by tugging hard on the line, after which it can be retrieved for a few yards in a series of short tugs. This will mimic the tail-flip escape response of the crayfish. When approached by a predator, the crayfish darts backward at great speed over a distance of 8–12 in, before swimming away at around 2 in per second by repeatedly flipping its tail.

MISCELLANEOUS ORGANISMS

THE NATURALS

A spider eats a mayfly trapped in its web but a trout may be next in the food chain.

Although insects are the major group of organisms that feature as trout food and therefore as successful models for artificial flies, a number of other animals must be mentioned.

Most animal species on Earth belong to a single very large phylum known as the Arthropoda. These creatures share many features such as the possession of jointed legs and a tough protective carapace or exoskeleton. Within the Arthropoda there are further large subdivisions. The insects and crustaceans have already been mentioned. Around 75,000 species of scorpions, harvestmen, ticks, mites, spiders, and a number of other small groups make up the Arachnida. Like insects, arachnids are important components of all terrestrial ecosystems. Unlike insects, they have eight not six pairs of jointed limbs, do not have wings, and are mostly

This water spider makes a nest of air held by a silken cradle to underwater plants.

carnivorous. A variety of spider and mite species are found in the vicinity of water habitats and are specifically adapted for a primarily aquatic life. Additionally many terrestrial spiders are at the mercy of the air currents as they balloon from place to place on silken threads and may end up on the water surface.

The Mollusca, another large group of invertebrate animals, includes familiar creatures such as land, freshwater, and marine snails, squid, and octopods. With up to 100,000 species the majority of these species are aquatic and marine. The biggest mollusk group, the Gastropoda, contains, among

Ramshorn snails have characteristically flattened, spiral shells.

many marine forms, the land and freshwater snails as well as slugs. They are readily recognized by their possession of a muscular foot and, usually, but not always, a hard shell which covers the body. Snails have a special file-like structure called the radula, which is used to graze plant material. Despite their protective shell, snails and their eggs are readily eaten by trout. Freshwater snails can be found grazing on the bottom, on plants or floating under the surface film. The Great Ramshorn Snail is one of a number of common and widespread species belonging to the genus *Planorbis*.

Closely related to the more familiar earthworms, leeches are aquatic annelids with suckers at the front and back ends. Around 1,000 species are known and they can be found in

Great Pond Snails can sometimes be carnivorous and will eat newts or small fish.

Leeches have suckers at either end of their bodies to hold on to prey and to anchor themselves in the water currents.

The tadpole of a bullfrog.

terrestrial, marine or freshwater habitats. Many species are predacious, feeding on snails, worms, and insect larvae, but more species are bloodsuckers being ectoparasites on marine or freshwater fish, birds and other vertebrates as well as invertebrates such as snails, crustaceans and insects.

Hungry trout do not just confine themselves to invertebrate provisions. At certain times of the year there may be an abundance of tempting, small, vertebrate food items. Although trout do not seem to like eating tadpoles, tadpole-like lures seem to work. It may be that such lures look like something else to the fish. Small fish are certainly favorite

A minnow, the smallest species belonging
to the carp family.

THE ARTIFICIAL ORGANISMS

ANGLERS' NAME
SPIDER

DRESSING	
HOOK: TMC 100, size 12	**RIB:** Silver tinsel with a tip at rear end
THREAD: Brown	
BODY: Brown dubbing blend	**HACKLE:** Brown partridge

FISHING METHOD
As for Lane's Wolf Spider.

ANGLERS' NAME
LARGE RED MITE

DRESSING	
HOOK: TMC 100, size 16	**BODY:** Red wool
THREAD: Red	**HACKLE:** Red cock

FISHING METHOD
Bright red mites (related to spiders) are sometimes common locally and can be found on the surface of the water. Allow to float along on the surface, imparting little if any movement.

ANGLERS' NAME
LANE'S WOLF SPIDER

DRESSING

HOOK: TMC 100, size 10

THREAD: Brown

BODY: Cork-brown silk tapered at each end

THORAX: Brown ostrich herl

HACKLE: Brown partridge tied between body and thorax

FISHING METHOD

There are two types of spider that might end up in a trout's stomach: those which live on and under the water (water spiders) and those which have ended up there by accident and drown. Fish the pattern by either moving the fly in a staccato but co-ordinated manner near the surface or letting it drift along, twitching.

ANGLERS' NAME
PARTRIDGE AND PEACOCK SPIDER

DRESSING

HOOK: TMC 100, size 12

THREAD: Black

BODY: Bronze peacock herl

HACKLE: Brown speckled partridge

FISHING METHOD

As for Lane's Wolf Spider.

FLOATING FRY

DRESSING

HOOK: TMC 300, size 6–10

THREAD: White for first part then black to finish

BODY: Underbody—fish-shaped tube of Plastazote with a red floss lateral line and throat; Overbody—pearl Mylar tubing

BACK AND TAIL: Peacock herl interspersed with black Twinkle

HEAD: Black thread built up to proportion

EYE: Self-adhesive decal covered in epoxy resin

FISHING METHOD

Either fish this pattern on a floating line to represent a stunned or dead fish, or fish it on a sinking line such that it

swims along above the bottom, alternatively rising and diving when the line is strip-retrieved at about 8 in per second, or swimming parallel to the bottom if retrieved smoothly.

PERCH FRY

DRESSING

HOOK: TMC 300, size 8

THREAD: Black

BODY: Gold tinsel with 2 short, dyed grizzle saddle hackles tied

in behind head

TAIL: Golden pheasant tippets

BEARD HACKLE: Red cock

FISHING METHOD

This pattern provides an authentic representation of a small perch. Fish so that it swims close to weeds and other submerged vegetation.

ANGLERS' NAME
ZONKER

DRESSING

HOOK: TMC 300, size 8
THREAD: Black
BODY: Silver Mylar dressed over wool gray rabbit Zonker

strip tied in at tail and head and glued to body
BEARD HACKLE: Yellow cock

FISHING METHOD

A lively imitation of a small fish which is best fished either deep or in midwater (depending on where the real bait fish

are lying). Retrieve either smoothly or to mimic the darting escape of a small fish. The Zonker strip wiggles and undulates realistically when wet.

ANGLERS' NAME
POLYSTICKLE

DRESSING

HOOK: TMC 300, size 6
THREAD: Black
BACK AND TAIL: Brown orange or yellow raffine
BODY: Black silk wound on to a silvered hook and crimson floss

wound on to hook at front end. Body then built into fish shape with polyethylene
HEAD: Varnished tying silk
BEARD HACKLE: Red or orange hackle fibers

FISHING METHOD
Another excellent minnow or other small fish imitation. Use either floating or swimming.

TADPOLE STREAMER

DRESSING

HOOK: TMC 5262, size 10

THREAD: Black

BODY: Black chenille tied flat over copper wire underbody

WINGS AND GILLS: Two black hen hackles tied in back to back at the tail

HACKLE: Sparse black hen at rear of body

FISHING METHOD

Trout don't actually seem to much like eating tadpoles, but tadpole imitations do catch fish. Fish the fly deep and retrieve

it in a series of short pulls of about ¾ in. Then let it sink down before re-commencing the retrieve. Tadpoles swim in short bursts before settling onto the bottom where they grub around feeding on algae.

TADPOLLY

DRESSING

HOOK: TMC 100, size 12

THREAD: Black

TAIL: Black cock or marabou

BODY: Ball of black dubbing blend tied at front of shank and covered with bronze peacock herl

FISHING METHOD

Another lifelike tadpole imitation. Fish deep as for Tadpole Streamer (above).

LEECH (POODLE)

DRESSING

HOOK: TMC 300, size 6

THREAD: Black

TAIL: Black marabou

TAG: Two or three turns of black wool

BODY: Black chenille

BODY PLUMES: Four or five small shuttlecocks of black marabou tied along top of shank

FISHING METHOD

Leeches swim by undulating their body up and down, a movement which is captured perfectly by the marabou plumes and tail in this pattern. Fish along the bottom, retrieving in short (¾ in) pulls at an overall rate of about 2 in per second.

FLOATING SNAIL

DRESSING

HOOK: TMC 100, size 10

THREAD: Black

BODY: Black or brown chenille rolled into a ball

FISHING METHOD

Aquatic snails of various species live on the bottom, among submerged vegetation, and suspended upside down from the surface film where they eat floating weed and algae. This pattern represents the last of these and should be fished in the surface film. Move the artificial fly very slowly and gradually.

SUGGESTED READING

Aquatic Insects. D. D. Williams & B. W. Feltmate (1992). C.A.B. International.

Aquatic Entomology: The Fisherman's and Ecologist's Illustrated Guide to Insects and their Relatives. W. P. McCafferty (1981). Science Books International.

Flies for Trout. Dick Stewart and Farrow Allen (1993). Mountain Pond Publishing.

Flies, the Best 1000. Randle Scott Stetzer (1992). Frank Amato Publications.

A Guide to Aquatic Trout Food. D. Whitlock (1994). Lyons & Burford Publishers.

Hatches II. Al Caucii and Bob Nastasi (1986). Lyons & Burford Publishers.

John Goddard's Waterside Guide (1991). CollinsWillow.

The Orvis Guide to Trout Stream Insects. Dick Pobst (1990). Lyons & Burford Publishers.

ASSOCIATIONS

Federation of Fly Fishers, P.O. Box 1595, Bozeman, MT 59771.

Trout Unlimited, 1500 Wilson Blvd., Arlington, VA 22209.

Salmon & Trout Association, Fishmonger's Hall, London, EC4R 9EL, U.K.

SUPPLIERS

L. L. Bean, Freeport, ME 04033.

The Orvis Company, Historic Route 7A, Manchester, VT 05254.

Kaufmann's Streamborn, 8861 SW Commercial Street, Tigard, OR 97223.

Blue Ribbon Flies, Box 1037, West Yellowstone, MT 59758.

Orvis Company Inc., 27 Sackville Street, London W1X 1DA, U.K.

Farlow's, 5 Pall Mall, London SW1, U.K.

Rothwell Rods, Reels & Flies, 29 Evison Road, Rothwell, Northants NN14 6AL, U.K.

ACKNOWLEDGMENTS

The Authors and Publishers are grateful to Dave Hughes of
Rothwell Rods, Reels & Flies for tying and helping to select the
flies; to Peter Gathercole for his advice, and expertise in the final
fly selection and for the fly-tying sequence; to Mr. Tom
Rosenbauer of Orvis Company, Inc., Manchester, Vermont for his
generous donation of dry flies.

PICTURE CREDITS

INDEX